For *An Intelligent Pers*

'Hugely enjoyable'

'A lucid guide to modern genetics' *Sunday Telegraph*

'Deeply provocative' *Business Times*

'A whirlwind exposition of evolution, the history of molecular biology, molecular genetics and the future' *British Medical Journal*

'Intelligently provocative' *New Scientist*

'Reading this book feels like having a conversation over dinner with a cultured, witty and well-informed companion' *The Lancet*

'An insightful tour of the history of genetics'
San Diego Union Tribune

'A deeply provocative read' *The Business Times*

'Elegant, impressive, sophisticated and lucid' *The Weekly Standard*

For *Life Without Genes*:

'Playful, filled with important and fascinating ideas, as well as entertaining metaphors'   *Times Literary Supplement*

'Readable and absorbing'   A C Grayling, *Financial Times*

'Crammed with lucid and fearless speculation on the origin and future of life'   *Independent on Sunday*

'Fascinating and bold ... interesting and profound'   *New Statesman*

'Gloriously playful, eye-opening and heartening. Woolfson wears his daunting erudition lightly'   *Scotsman*

'Woolfson writes beautifully and both entertains and informs'   *Times Higher Education Supplement*

'Wildly, ferally enjoyable'   *Scotland on Sunday*

'Woolfson is a virtuoso in full command of extraordinary material'   Simon Ings

'Serious fun'   Graham Cairns-Smith

'One of the most exciting reads in years'   *Cambridge Evening News*

'Yields many insights ... much to mull over'   *The Irish Times*

'Anyone who has ever wondered how a planet of rocks and boiling seas could have given birth to Mozart and the Spice Girls will find the answer here'   *Manchester Evening News*

'A pleasure to read'   Ian Stewart

An Intelligent Person's Guide to Genetics

Adrian Woolfson

Duckworth Overlook
London • New York • Woodstock

This edition 2005
First published in 2004 by
Duckworth Overlook

LONDON
90-93 Cowcross Street
London EC1M 6BF
inquiries@duckworth-publishers.co.uk
www.ducknet.co.uk

NEW YORK
The Overlook Press
141 Wooster Street
New York, NY 10012

WOODSTOCK
The Overlook Press
One Overlook Drive
Woodstock, NY 12498
www.overlookpress.com
[for individual orders and bulk sales in the United States,
please contact our Woodstock office]

© 2004 by Adrian Woolfson

The right of Adrian Woolfson to be identified as the
Author of the Work has been asserted by him in accordance
with the Copyright, Designs and Patents Act 1988.
All rights reserved. No part of this publication
may be reproduced, stored in a retrieval system, or
transmitted, in any form or by any means, electronic,
mechanical, photocopying, recording or otherwise,
without the prior permission of the publisher.

A CIP catalogue record for this book is available from the
British Library and the Library of Congress

ISBN 0 7156 3423 2 (UK)
ISBN 1-58567-810-4 (US)

Typeset by Ray Davies
Printed and bound in Great Britain by
CPD Ltd, Wales

Contents

Preface	7
1. The DNA time machine	9
2. How to build a living thing	29
3. Fabulous monstrosity	51
4. Smart genes	75
5. Reprogramming life	105
6. Making creatures from scratch	125
7. The limits of possibility	146
8. A manifesto for life	177
Bibliography	209
Index	230

To Claire and my parents

Preface

The sight of a miniature ship perfectly assembled inside a glass bottle can impress even the deepest sceptic. But mystery always fades when we learn the simple tricks that underlie apparently remarkable feats. The creation of living things from inanimate matter is no different. The tricks in this case lie in the DNA sequences that program their construction. They teach us how a crocodile can be transformed into a kangaroo, and how the tiniest and most magnificent examples of life are constructed. Together, these tricks define the substance of genetics. Genetics does not disprove the existence of God, but it does teach us that every aspect of our appearance and behaviour can in some way be accounted for by the operations of our body. It also demonstrates that we could, if we wished, synthesize new forms of life from first principles, and change ourselves beyond all recognition. The twentieth century outlined the key principles of genetics; in the twenty-first, this knowledge will transform human existence in unimaginable ways.

In the eighteenth-century Enlightenment, European civilization was transformed from a consciousness dominated by religious doctrine into its modern secular incarnation. We are at the cusp of a new Enlightenment, defined by the accumulated genetic knowledge that enables us to entertain the possibility of modifying our own nature and of creating artificial life. This tremendous power for change is unprecedented. This book is an attempt to account for that power by examining the inner workings of genetics. It should not be read as advocating a particular course of action. My view is slightly different: the creation of synthetic life is an inevitability.

Cambridge, August 2004.

1

The DNA time machine

Imagine yourself transported back to the streets of Victorian London. Hovering high over the banks of the Thames, you peer down at the figures and activities below. These characters are long gone, the business of their lives forgotten. All that remains are their shadows, glimpsed through the writings of authors who documented the comings and goings of everyday life. Unable to transport ourselves back in the flesh, we need a guide: in this case, James Greenwood, whose *Journeys through London* (1867) is a conduit into an unfossilized episode of human affairs, helping us re-create a snapshot of past activities on the Thames. As we look down, we see the 'ragged, tailless coat, minus the sleeves' and the 'wonderful collection of materials and colours fashioned somewhat to the shape of trousers' that was the uniform of the impoverished 'gleaners' or 'mud-larks' who made a living sifting through riverbank mud in search of submerged treasures. Further on, we see the gas-lit cobblestones of Brick Lane, wicker baskets, the veal and mutton-chops of the street-sellers at Newgate and the precious metal and diamond stalls of Hounsditch. Greenwood noted that the mud-larks often had 'the "luck" to turn up something curious should a party looking at all like a buyer be in their vicinity'. A friend of his 'happened to be walking near one of these treasure hunters, just in time to see him turn up a rare old apostles' spoon, very massive, and still bearing faint traces of the goldsmith's mark'.

Such vignettes are rare: most of history has gone unrecorded. The dialects, mannerisms, pastimes and preoccupations of countless individuals, cultures and civilizations have vanished without trace. But there are glimpses. Terracotta fragments, carvings and cave daubings offer tantalizing insights into the lives of our earliest ancestors. More recent records – artefacts, written accounts, photographs, sound recordings and film and television footage – enable us to see Punjabi maharajas or soldiers emerging from the trenches in Flanders. Like the Roman coins dug up by the mud-larks, eyewitness and cinematographic accounts are part of our common cultural treasure, teaching us not to be complacent about the commonplace, and to respect life's minutiae. The modern world would be incomprehensible to the author of *Mixing in Society*, who wondered in 1874 if anyone would 'venture to predict the programme of an evening party in the year of grace 1969'. Our own pastimes will one day seem as strange as those of the mud-larks. Even the most mundane aspects of day-to-day life, such as a visit to the supermarket, will appear extraordinary.

But the history of life is ancient, far older than the affairs of mankind. The unwitnessed events of prehistory are not completely opaque, however. All life, past and

1. THE DNA TIME MACHINE

present, is linked by a common record written in the chemical code from which genetic material is made. DNA consists of long sequences of four chemical 'letters' – C, T, G and A – strung together in different combinations like differently coloured beads on a necklace. The information of DNA is encoded in the precise order of these four chemicals – like writing, but using fewer symbols. To an archivist of past events, the DNA sequences of extinct and living things are precious, preserving the record of many aspects of the history of life on Earth – in particular its prodigious array of structures, from starfish to sabre-toothed tigers. Each sequence has a unique story, the distillation of a relentless journey through time and circumstance. Together, they comprise a patchwork of life's history stretching back to its origins. But the pieces of this fabric are dissociated, and some are lost: they must be pieced together with careful detective work.

Most natural phenomena – the vortex that emerges spontaneously when a plug is pulled from a bathtub, or the self-organizing geometry of snowflakes – leave no trace behind. Living things are unique in containing an internal record of themselves. The details of a creature's form and aspects of its behaviour are computed from the coded information inside the tiny threads of DNA, which function as miniature instruction manuals for its construction and operation. The DNA database controls all events in living things, but only the most essential information is recorded. So although the sequence of a DNA fragment rescued from the clothes of a nineteenth-century chimney-sweep could tell us much about him, it would not communicate his individual habits, dialect or lifestyle. Such things are not recorded in DNA, being transmitted by social learning or lost forever. If a chimney-sweep could one day be re-created from pieces of DNA, he would speak in the vernacular of current times rather than in Victorian London slang. Genes encode the raw potential for speech, behaviour and culture, but in 'higher' organisms their details are often filled in by experience. And though a songbird is born with the basic ability to sing, the songs it overhears in its youth determine their precise nature. As a result, birds living in inner city areas have been known to produce song repertoires that sound like car alarms. The extent to which genes constrain the cultural attributes of humans and other animals, however, remains unclear.

In July 1918 Russia's royal family, the Romanovs, were taken into the basement of their summer dacha and shot by Bolshevik revolutionaries. Their bodies were removed in a lorry. The final resting place of Tsar Nicholas II, the Tsarina, five children and their servants remained a mystery for many years, until an eyewitness account found in Communist Party archives indicated that they had been buried in woods near Ekaterinburg. Although the bodies of the Tsarevich, Alexei, and one of the princesses were never found, analysis of a collection of skeletons found in a shallow grave in the region indicated that they were the remains of the Romanovs. This was inferred by comparing DNA taken from the skeletons

1. THE DNA TIME MACHINE

with that of existing and deceased Romanov family members and descendants, including the Duke of Edinburgh and the Tsar's 28-year-old brother, the Grand Duke Georgij Romanov, who died of tuberculosis in 1899. Although unable to deliver the precision of an eyewitness account, the DNA record in this way helps reconstruct the last moments of the Romanov dynasty. It could also yield other information such as the Romanovs' disease predispositions, and might one day help reconstruct them.

DNA is unstable, and sensitive to destruction by enzymes, oxygen, water and sunlight. But it is possible to recover DNA much older than that of the Romanovs. The first ancient DNA was extracted in 1984 from a preserved quagga in a German museum. Quaggas were zebra-like creatures which once populated the plains of southern Africa but became extinct in the nineteenth century following extensive hunting. The last individual died in an Amsterdam zoo in 1883. They differed from zebras in that they had stripes only on their head, neck and the front of their body; the rest of their skin was brown, with a white underbelly and legs. All that remains of this once abundant creature are twenty-three mounted skins housed in museums around the world and a few articulated skeletons. DNA analysis has shown that rather than being a unique species, quaggas were a subspecies of the zebra. If they once interbred with zebras, many of their genes are likely to persist in modern zebra populations. With the help of careful breeding, it might be possible to reassemble some of these genes in a single creature – and in so doing resurrect the quagga. Whether it would be reasonable to call such hypothetical resurrected creatures quaggas, however, is another question. Despite having a superficial anatomical resemblance, they would differ from historic quaggas in many ways. It would be hard, for example, to know whether the behaviour and internal structure of a resurrected quagga resembled that of its historic counterpart. As there are no living examples for comparison the issue would be practically impossible to resolve.

Since the isolation of material from the quagga, DNA from far older samples has been retrieved. This includes fragments extracted from human remains taken from a house in Pompeii dating to the eruption of Vesuvius in AD 79, a 2400-year-old Egyptian mummy and a 5000-year-old Ice Man found in the Tyrolean Alps. And, most interestingly perhaps, from the remains of a dodo stored in the Museum of Natural History in Oxford. The material used for extraction was taken from the only existing dodo flesh in the world, from a specimen brought to Europe for exhibition in the second half of the seventeenth century. A routine clear-up in the 1700s had nearly consigned the remains to the flames, but a quick-thinking curator rescued the head, leg and foot. These remains eventually became known as the 'Alice in Wonderland dodo': the mathematician and author of *Alice in Wonderland*, Charles Dodgson, once took Alice Liddell, the model for Alice in his book, to see it. The memory led him to include a dodo in his story.

Dodos were large, flightless birds which once inhabited

1. THE DNA TIME MACHINE

the island of Mauritius, where they foraged on the forest floor. Their unpalatable meat ensured that the Dutch sailors who first noticed them in 1598 largely ignored them. But feral pigs, dogs and rats introduced by the colonizers ate their eggs and led to their demise. The last confirmed dodo sighting was in 1662 by a sailor called Volkert Evertsz on an islet off Mauritius. The actual extinction may have been slightly later: a runaway slave claimed to have seen one in 1674. Their distinctive appearance gave few clues as to their origin: it was a big surprise when DNA analysis showed that dodos were related to pigeons. This indicates that dodo genes might lurk in pigeons to this day. Like the quagga genes dispersed in zebra populations, genes rescued by intensive breeding of the Nicobar pigeons of Southeast Asia, to which dodos are most closely related, might one day help resurrect the extinct species. But it is not only studies of extinct creatures that have delivered unexpected results. DNA comparisons between living creatures can be equally surprising. The marsupial wolf of Australia, thought to be closely related to the carnivorous marsupials of South America, is in fact unrelated. The shared features result from evolutionary convergence on the same design, rather than a shared family tree.

Most of the DNA obtained from ancient specimens is degraded into small pieces and contaminated with DNA from other species, such as bacteria and fungi. Investigating archaeologists, anthropologists and biologists are themselves often sources of contamination. As a result, ancient DNA must be prepared with great caution. Animals have two types of DNA, which differ in their size and storage site. Nuclear DNA contains the bulk of the genetic archive, including the parts needed to make living things. It is stored in the cell nucleus, a tiny compartment in the middle of the cell. Mitochondrial DNA, on the other hand, contains a very small and specialized set of genes involved in energy production. These are housed in minute cytoplasmic organelles called mitochondria. Each cell has only a single copy of nuclear DNA but hundreds of copies of mitochondrial DNA, reflecting the large number of mitochondria a cell contains. Since there are more copies of mitochondrial DNA than nuclear DNA, DNA recovered from ancient specimens is usually mitochondrial in origin.

DNA is prone to damage from day-to-day wear and tear and the occasional mistakes that occur when it replicates itself. Like railway workers fixing train tracks, specialized protein machines rapidly repair such damage. But not all damaged DNA is repaired to the same extent. Nuclear DNA repairs damage to its sequences with great precision, but repair of damaged mitochondrial DNA is less efficient. As a result, mitochondrial DNA sequences change more rapidly over time. Changes to DNA sequences are known as mutations. Most mutations occur at a single position in the DNA sequence: these are known as point mutations. Alternatively, much larger changes can be introduced, with bits of DNA being cut and pasted into one another: this process is called recombination.

1. THE DNA TIME MACHINE

Modifications to DNA affect the nature, the cellular location, and the time and rate of production of the protein molecules from which living things are constructed; they provide the raw material for evolution to act on. Even the smallest changes can result in significant changes to the structure and function of a living creature.

Unlike nuclear DNA, which is inherited from both parents and is subject to reshuffling, mitochondrial DNA is inherited maternally and provides an uncorrupted genealogical record. The fact that it mutates rapidly over time but is not corrupted by reshuffling means that mitochondrial DNA offers a unique window into the recent past, including the history of mankind. Through this window, many aspects of life's history – including evolutionary relationships, population migrations, ecological structures, animal behaviour and diet, and aspects of the early human mind – can be indirectly viewed. But in order to read this record it is necessary to decipher the DNA molecules from which the historic archive is built. The extraction of DNA from bones and preserved tissues is straightforward and these days quite routine. When cells are broken open with detergents, DNA goes into solution. It can then be precipitated out and becomes visible as a white, shiny, gelatinous mass. When only a small amount of DNA can be extracted, a technique known as the polymerase chain reaction (PCR) can be used to amplify it. This method is so powerful that it can detect DNA from a single cell, or even a single fragment of DNA from a single cell. Like the radio telescopes that enable astronomers to peer back into the history of the universe, PCR has revolutionized the study of ancient DNA and opened up new vistas of life's past.

Once the target DNA has been amplified, the sequence of the four chemical 'letters' from which DNA is made must be decoded, using a technique pioneered in the 1970s by the Cambridge scientist Fred Sanger. Combinations of these letters in 'coding regions' represent the genetic information needed to make proteins. 'Non-coding regions' contain the information needed to regulate the time, place and pattern in which genes are switched on and off. A single gene consists of both the coding region and the associated non-coding regions, which function as a control unit or 'mini-brain' that regulates the gene's behaviour. But the concept of a gene is in reality a little more complex. This is because whereas the regions encoding proteins are located together, the non-coding regions that control them may be scattered over great distances. The control regions of DNA also integrate genes into the regulatory networks that co-ordinate the metabolism of cells. So although the coding regions of genes are easily identified, it is often hard to say where the gene in the greater sense begins and ends. DNA comes in long, continuous pieces that are bundled up to form chromosomes. Rather than having one very long chromosome, organisms usually have multiple chromosomes which together contain the core information needed to build and operate the creature in question. New, efficient methods for DNA sequencing make it possible to sequence an organism's entire DNA archive, both coding and non-coding. The complete DNA sequence housed in a cell is known as its genome; complete genome sequences have been determined for an ever-increasing collection of organisms, including viruses, bacteria, yeasts, plants, worms, flies, fish, mice, rats, monkeys and humans.

The extraction of mitochondrial DNA from an ancient bone found in Germany in 1856 enabled the first observation of the genes of ancient humans and helped to resolve a central question of early human history. Neanderthals, named after the Neander Valley, where the first bones were found, were a short and stocky type of human with large heads and trunks which became extinct around 30,000 years ago. Their significance in mankind's history was disputed for many years. Some scientists believed that Neanderthals were our direct ancestors; others saw them as a separate species of human that did not interbreed with our ancestors. If this were the case, their genes would not have contributed to building modern humans. Analysis of mitochondrial DNA recovered from Neanderthal bones has shown that our direct ancestors were in fact a separate species, which emigrated out of Africa and resembled us more closely than Neanderthals. Neanderthals, it turns out, seem to have been an irrelevant side branch of the human evolutionary tree which eventually came to an abrupt end. The African ancestors of modern humans and Neanderthals evolved from a common ancestor and co-existed in Europe for thousands of years, but the species did not interbreed. So apart from those resurrected from old bones, Neanderthal genes have been lost forever. The reason for the Neanderthals' extinction is unclear. But since they became extinct within 15,000 years of the arrival of the modern-looking African immigrants, it has been suggested that they were the victims of a genocide perpetrated by our ancestors.

With a little ingenuity, DNA analysis is also able to provide detail at a much finer level. Without an eyewitness to observe what extinct creatures ate, one might think that this and other aspects of their behaviour would have been lost forever. But the engine of the DNA time machine may once again be of assistance, squeezing history out from the most intractable artefacts. Analysis of DNA from the droppings of extinct animals enables changes in eating behaviour and aspects of the surrounding environment to be inferred. The faecal pellets of ground sloths living in southern Nevada 28,500 years ago, for example, contain the DNA of pine needles, whereas those living in the same region 20,000 years ago do not. This suggests that pine forests had disappeared in this region by that time; the presence in the droppings of DNA from contemporary plants shows that their diet had adapted to the environmental changes. Similarly, analysis of faecal samples from 2000-year-old Native Americans shows that their diet included at least eight types of plant, as well as antelopes and bighorn sheep.

DNA analysis can take us even deeper, however, enabling transportation into the minds of our ancient ancestors. Ancient cave paintings present dual mysteries for modern anthropologists. What were the paints made from, and what significance did the murals have for the artists? Were they purely decorative or did they have a deeper significance? The DNA time engine allows an imperfect forensic insight into our ancestors' thoughts. Inspecting 3000- to 4000-year-old rock paintings from Texas, the chemist Marvin Rowe noticed that the paintings were covered with mineral deposits of calcium carbonate and speculated that these might have protected the DNA present in the pigments. This turned out to be the case, and he was able to isolate DNA from paint fragments. Analysis indicated the presence of DNA thought to be from either a deer or a bison. If the DNA were indeed from a bison, that would be a very significant discovery, since bisons were not at the time indigenous to that region. The finding suggests that the artists had gone to great lengths to obtain their pigments rather than using readily available resources, demonstrating that the paintings must have had a special significance and that their creation was motivated by an underlying belief system. Ancient humans, it seems, may have resembled us more closely than we imagined, not only looking like us but thinking and behaving like us too.

The ability of the DNA time machine to take us back more than 100,000 years appeared to be limited by the inevitability of DNA's decay. And those 100,000 years represent only a fraction of life's three and a half billion or so years of history. So is it possible to travel beyond this? Studies of ancient DNA found in the frozen permafrost of Siberia, where the extremely low temperatures have preserved the material, show that it is possible to rescue DNA up to two million years old. Although the remains of plants and extinct animals such as mammoths and steppe bison that once inhabited the permafrost are scarce, DNA has been isolated from permafrost sediments devoid of fossils. This disembodied permafrost genetic archive releases us from the usual need for fossils. Plant DNA found in Siberia is thought to have originated from ancient roots; that of mammoths and other animals is likely to have originated from cells excreted in urine and faeces. The extraction and analysis of DNA from such sources has enabled a partial reconstruction of this region's ancient ecology, flora and fauna. Rather than being the sparse and lifeless tundra we assumed, the Beringia region – which spanned eastern Siberia and western Alaska – turns out to have been a vegetation-rich steppe. The fact that grasses declined from 36% to 3% in all the DNA samples from around 11,000 years ago supports the idea that climate change was responsible for the mass extinction of mammoths and other Siberian plants and animals. DNA analysis of the glacial vegetation can also help test computer models of climate changes that occurred around that time.

But even if the upper limit for DNA survival were extended to two million years, our DNA time machine would still be impotent with respect to the earliest moments of life's history, hundreds of millions and ultimately over three billion years ago. But there are tricks that can take us closer, enabling us to reconstruct plausible candidates for the DNA sequence of ancient creatures. Whereas the study of human origins requires a focus on relatively recent history and rapidly changing mitochondrial DNA, excursions into life's deepest history of life demands a different level of resolution, and the use of nuclear DNA, which changes only slowly over time. In order to travel as far back as possible, however, a new conceptual method is needed. The approach is similar to that used by linguists to trace the origin of modern languages from an ancestral mother tongue; it involves the construction of genealogical trees and the tracing of DNA lineages. Once a family tree of DNA sequences from a group of related species has been established, it can be used to infer the extinct DNA sequences of the ancestor species at the root of the tree from which species on later branches diverged. With a little further detective work, inferred sequences can also help reconstruct aspects of their lifestyles and the physical environment that these creatures inhabited.

In 1963 the biochemist Linus Pauling predicted that it would one day be possible to use molecular 'restoration studies' to reconstruct extinct forms of life. If a cluster of related DNA sequences could be used to create a family tree, then the DNA sequences of the common ancestor at the root of the tree could be inferred from the sequences of descendant species. Once the DNA sequence of an ancestral gene had been reconstructed, the gene could be 'brought back to life' by artificially synthesizing the protein that the gene encodes. This would be achieved by inserting the gene into another organism, which would subsequently treat the gene as one of its own, using its instructions to make the encoded protein. The resurrected protein could then be purified and subjected to tests. If this could be done on a large scale and all the genes of a hypothetical extinct common ancestor inferred, it might be possible to bring the ancient creature back to life. The chronometer on the DNA time machine can be roughly calibrated from knowledge of the rate at which changes are fixed into DNA sequences, enabling the DNA time traveller to determine the approximate epoch of history under view. The method has a drawback, however: the reconstructed sequences must always remain tentative, and will never be the exact sequences of the original creatures.

Imagine now that once again we are hovering high over the banks of the Thames. This time there is no sign of the smoking Victorian city or the mud-larks sifting through the clay. All we see is vegetation, devoid of humans or animal life. The chronometer on the time machine reads 600 million years before the present. But our view is clouded. It is unclear whether the landscape is covered with snow, or whether hot springs bubble beneath us. To clarify the scene, we need further assistance, in this case provided by Eric Gaucher and his colleagues at the University of Florida, whose work enables the physical environment of the time to be inferred. They assembled a collection of DNA sequences from around fifty different

1. THE DNA TIME MACHINE

bacterial species, each encoding the same 'elongation factor' – a protein essential in protein synthesis. Elongation factors are so essential to life that they change almost imperceptibly over time, making them perfect for such studies. With their help, Gaucher's team constructed an evolutionary tree for the bacteria in their sample. This enabled them to infer the DNA sequence of an ancestral elongation factor gene that was present 600 million years ago in a hypothetical common ancestral bacterium. It was then a trivial thing to introduce the reconstructed gene into modern bacteria, which used it to synthesize the ancient elongation factor protein. The resurrected protein was then subjected to various tests. It was found to function optimally at a temperature of between $55°C$ and $65°C$, unlike the majority of modern elongation factors, which work best between $20°C$ and $40°C$. These studies will need much checking before they can be definitively relied on, but they suggest that rather than being snow-lined or filled with boiling springs, the landscape 600 million years ago was somewhere on the hotter side of the midpoint of these two extremes.

Time travel to such distant epochs can be achieved by other means as well. One method involves the comparison of DNA sequences from different species. Multicellular life is known to have originated in a geological era known as the Precambrian, around the period of the Gaucher dating. At around this time the thirty-two or so basic animal body plans, or phyla, which are the basic design templates for all living things, first appeared in the fossil record. One of these phyla, chordates, gave rise to all vertebrates, including mice and humans. Other phyla include arthropods, whose body plan is shared by all insects, including houseflies; and annelids, which include worms. Animal phyla such as sponges and corals appear to have very little in common with humans. It was therefore surprising when a recent study of coral showed that it has a large cache of genes in common with humans. Furthermore, 10% of the genes found in humans, including genes encoding parts of the nervous system, were not found in flies and worms, which evolved millions of years after coral. This dramatic finding suggests that some of the genes thought to be 'vertebrate innovations' in fact evolved much earlier. It also indicates that some animal species, such as worms and flies, actually discard genes as they become more complex. Studies of coral tell us more about certain aspects of animal evolution and our own humanity than worms and flies – whose diversification involved the loss of some of the genes that help build humans.

Once the genomes of all existing and all possible extinct species have been sequenced, it should eventually be possible to reconstruct life's complete evolutionary tree. Through careful study of the nodes and branches of this tangled web of DNA sequences, it will be possible to reconstruct extinct ancestral species with increasingly greater certainty. Each time an existing species is destroyed by extinction, a degree of resolution in the reconstruction is lost. The surprise findings in coral demonstrate that crucial information may be found in the

1. THE DNA TIME MACHINE

most unexpected places – so it is essential that all life is equally respected, however small, obscure or apparently uninteresting. Eventually it might be possible to re-create the elusive ancestor of all life on Earth, a hypothetical organism known as LUCA, or the 'last universal common ancestor'. As with the remnant dodo genes found in pigeons and the quagga genes in zebras, the remnants of LUCA should be scattered across the genomes of all living things. It is simply a matter of piecing together all the bits of information, and then generating candidate genomes which could be reconstructed, initially gene by gene and then in their entirety.

So far a comparison of genome sequences from a wide range of organisms has identified 60 genes that appear to be common to all living things and which are likely to have been present in LUCA's genome. But LUCA would have had many more genes than that, probably at least 600. LUCA's genes would be expected to reveal something about the world in which it lived, 3.6 billion years ago. LUCA must have been an extraordinary creature, containing within its small DNA database the potential for all life's subsequent history. Recently, however, the idea that LUCA was a single organism has been challenged by an alternative view: that LUCA was a last universal commune of promiscuous ancestral organisms which swapped genes with one another, muddling the root of life's genealogical tree. It is possible that such gene-swapping has eroded the earliest genealogical record and that LUCA may therefore never be knowable. But it should one day be possible to make some reasonable attempts at bringing a creature that approximates to LUCA, or to members of the LUCA community, back to life. In reconstructing the past, we will also have outlined an engine capable of travelling into the future.

2

How to build a living thing

Shortly after Charles Darwin died in 1882, Sir Richard Owen – one of the greatest comparative anatomists of the Victorian age and apparently the only man whom Darwin ever really disliked – received a letter from the Rt Hon. Spencer Horatio Walpole, a member of parliament and a trustee of the Natural History Museum in London. Walpole asked Owen whether he thought Darwin's accomplishments merited a statue in the Natural History Museum. Although Darwin had not managed to persuade everyone that different species had originated as the result of natural rather than metaphysical processes, he had made it acceptable to believe that this was the case. It was no longer possible to presume that life was created simultaneously in an immutable incarnation by an intelligent designer. But Darwin's success with *The Origin of Species* had in some ways come at the expense of biologists like Owen, who, though accepting that new species could emerge by natural means, insisted on an underlying divine Creation. Owen had spent two decades lobbying parliament for funds to build the Natural History Museum and had been closely involved with its design. The notion that he should support the construction of a statue to glorify his rival was not one he cherished. A carefully crafted response was required.

In his reply to Walpole on 5 November 1882, Owen began by remarking on the importance of Darwin's contributions to natural history, but went on to undermine him. He wrote that Darwin had solicited for the post of ship's naturalist on *HMS Beagle*, whereas he had in fact been invited to join the voyage as a companion for the depressive captain, Robert Fitzroy. And though Owen praised Darwin's work on coral-atoll formation, he ignored Darwin's other significant work, *The Descent of Man*, which outlines how humans might have evolved from primitive ancestors. Owen called Darwin the 'British Copernicus of Biology', but did not miss the opportunity to say that – like Copernicus, whose ideas were valuable to the newly emerging science of astronomy but were eventually overshadowed by Galileo, Newton and Kepler – Darwin's rudimentary exposition was also likely to be eclipsed. In a further swipe, he added that 'Darwin parallels Copernicus, save that the latter not only knew not, nor feigned to know, how the planets revolved around the sun'. Owen predicted a future in which the laws underpinning the evolution of new species would be 'as firmly established as the law of gravitation', and concluded that the question of whether 'the estimation of scientists at home or abroad of Charles Darwin's claims to posthumous honour be met, or their expectations fulfilled by placing a statue in the Museum of Natural History' was one for 'Administration'. Darwin did eventually get his statue, as did Owen, though they were diplomatically placed on different staircases.

There was a grain of truth in Owen's protestations. Although his theory of natural selection was essentially correct and was certainly a significant achievement, Darwin lacked any insight into the nature of the generative processes that created new species. His own speculative attempt at a theory of inheritance, known as 'pangenesis', was similar to that of the ancient Greek philosopher Hippocrates, and the reverse of contemporary genetic theory. Darwin proposed that every cell in the body produced tiny buds, or 'gemmules', encapsulating the experience gained in their lifetime. These experience-laden particles made their way to reproductive cells (where Darwin conceded there might be a problem with overcrowding). He was hampered by the lack of an experimental methodology through which the problem of the origin of species could be made tractable. Like almost all his contemporaries, he was unaware of the pea-breeding experiments of an obscure Augustinian monk, Gregor Mendel, experiments that would eventually form the cornerstone of modern genetics. Mendel's work was published in 1866, seven years after *The Origin of Species* but sixteen years before Darwin died. The mechanism of inheritance had been well within Darwin's grasp, but eluded him.

Mendel, the son of a peasant farmer, was born in 1822 in Heizendorf, a small village near the border of northern Moravia and Silesia. After joining the monastery of St Thomas in Brünn and being ordained as a priest in 1847, he went to work as a teacher in Znaim, but failed his teaching certificate examinations. He then enrolled as a trainee teacher of mathematics and biology at the University of Vienna, before returning to the monastery at Brünn in 1854. He had tried to retake his teaching certificate but withdrew because of illness. In 1856, with the support of the abbot, Cyrill Franz Napp, he began a series of simple hybrid cultivation experiments using the garden pea *Pisum sativum*. At the time, the town of Brünn was the centre of the Austro-Hungarian textile industry; the high cost of imported Spanish wool spurred civic leaders to foster and promote breeding enterprises that might improve local flocks. The context obviously stimulated him, but Mendel was mainly driven in his work by his fascination with ornamental plants: he liked the colours of the flowers and wanted to extend their variety.

In 1862 Mendel travelled to England to see the Great Exhibition, where he came across a German translation of Darwin's *The Origin of Species*. It seems to have affected him greatly, and given his religious beliefs he was presumably determined to prove Darwin wrong. Mendel focused on basic differences between the characteristics of peas, interbreeding plants with easily distinguished traits: red and white flowers, or smooth and wrinkled seeds. By carefully documenting the results of his crosses, he was able to discern distinct patterns in the way simple characteristics were transmitted from one generation to another. Mendel saw these recurring patterns of inheritance

2. HOW TO BUILD A LIVING THING

as evidence of the imprint of divinity, rather than the godless change invoked by natural selection.

Mendel's great misfortune was that he published his work, 'Versuche über Pflanzen-Hybride' ('Treatises on Plant Hybrids'), in an obscure journal. The publication of Volume 4 of the 1866 'Records of the Brünn Association for Natural Research' – *Verhandlungen des naturfoschenden Vereines in Brünn* – went largely unnoticed by the rest of the world. Botanists were not then familiar with the idea that mathematics could be used to study biological problems, and Mendel lacked the charisma to enthuse his colleagues. His presentations to the Association in February and March 1865 were met with indifference. The final blow came when Mendel requested that reprints of his article be sent to a broad selection of botanists and other scientific dignitaries. One of the forty recipients, Carl Wilhelm von Nägeli, an eminent German botanist, began a protracted correspondence with him. He urged Mendel to continue his work, but suggested he use the hawkweed *Hieracium* rather than peas. This proved to be very bad advice. The hawkweed's asexual reproduction meant that its genetics were entirely different from those of peas and quite unsuited to hybridization studies. The failure of the new experiments, and the work involved in running the monastery after he was elected abbot in 1868, led Mendel to abandon his work on plant hybrids, instead devoting his spare time to bee-keeping and meteorology.

The legacy of Mendel's simple experiments was enormous. He had created the first method for the systematic study of heredity and shown that the agents of heredity were discrete particulate units that came in different versions and led to variations in the characteristics they encoded. He also demonstrated that the transmission of characteristics across generations is governed by simple laws. Mendel's work was mentioned in an 1881 textbook but was not properly noticed until its simultaneous rediscovery in 1900, sixteen years after Mendel's death, by Hugo de Vries in Holland, Carl Correns in Germany and Erich von Tschermak in Austria – a rediscovery that marked the beginning of what has been called the 'century of the gene'. William Bateson, a Cambridge zoologist, played a significant role in the popularization of Mendel's laws, in part by translating Mendel's work into English. Bateson realized that a new word would be needed to unify the study of inheritance and variation, and in 1908 wrote that 'such a word is badly wanted and if it were desirable to coin one, Genetics might do'.

But it took an American professor, Thomas Hunt Morgan, who had previously worked on regeneration in worms and sea urchin development, to establish the physical reality of genes and to translate Mendel's ideas into a chromosomal theory of heredity. Before Morgan, chromosomes had been observed under a microscope in the nucleus of cells, but had never been linked to the transmission of genetic information. Morgan, who was initially sceptical about Darwin's theory, would eventually outline the physical basis of molecular evolution. He was born in Kentucky in 1866 and in 1904 took up the

2. HOW TO BUILD A LIVING THING

newly created chair of experimental zoology at Columbia University. More interested in animals than plants, he was keen to devise an experimental system to ascertain whether Mendel's principles applied to organisms other than peas. In 1907 he started work on the tiny fruit fly *Drosophila melanogaster*, so called because it feeds on decaying fruit. It was a good choice: flies were fertile all year round, producing new offspring every twelve days, and their diminutive size meant that thousands could be housed at once, using glass milk bottles. Morgan hoped that if he observed the flies over many generations he would eventually spot a mutant – a fly that had undergone a sudden and spontaneous change in its body form. Two years went by without his obtaining a single one. He was immensely disappointed, and complained to a laboratory visitor that he had wasted two years of his life. He persisted nevertheless, and in April 1910 he noticed a male fly with white eyes rather than the usual red. He had finally demonstrated that he had a system for studying the inheritance of a simple animal trait.

Morgan set up a 'Fly Room' in the zoology department and worked with his students at the laboratory bench to try to unpick the physical basis of the white-eyed mutant. Their tiny lab – Room 613 at Schermerhorn Hall in Columbia University, measuring sixteen feet by twenty-three and containing only eight desks – was to become a powerhouse that would establish genetics as the central paradigm and agenda of twentieth-century biology, in place of anatomy and armchair speculation. Using the white-eyed mutants and adopting the term 'gene' introduced by the Danish botanist Wilhelm Johannsen in 1909 in place of 'Mendel's factor', Morgan and his tightly-knit group of students showed that the genes encoding heritable traits lie at precise positions on chromosomes and are linked to form a linear chromosome map. The discoveries were made by mating the white-eyed male mutant with its red-eyed sister. All the offspring had red eyes, suggesting that the red colour was dominant. When the second generation of flies were mated, white-eyed individuals were seen again, but all of them were male, suggesting that eye colour was in some way linked to gender. When the flies' chromosomes were examined under a microscope, Morgan and his team were able to infer that the gene for eye colour was located on a specialized sex chromosome that determined both gender and eye colour. This was the first time a trait had been correlated with the presence or absence of a chromosome. Since the orderly assembly of thousands of such traits underlies the construction of living things, Morgan's work promised to yield the secrets underlying the origins and modification of all species. The fact that such traits as gender and eye colour occasionally became unlinked indicated that pieces of chromosomes could exchange positions, by a process that came to be known as recombination. Recombination brings together mutations that had not previously been located on the same chromosome, allowing new possibilities of variation. Chromosomes occur in pairs, so it seemed likely that the exchange occurred between identical regions of separate chromosomes from each pair. This suggested that genetic information could be modified in two separate ways: by mutation, and by the reshuffling of the linear position of genes, in a process of recombination.

Morgan summarized his findings in *The Mechanisms of Mendelian Heredity* (1915). He had given genes a physical reality and a precise location in the chromosomes, and had established them as the unifying factor that determined heredity, development and evolution. He had found a physical mechanism to explain the construction of living machines. The ancient doctrine of vitalism, which asserted that the structures of living things were dictated by non-specific 'life forces', could slowly be dismantled. Morgan's mutant flies also offered a simple experimental system for the resolution of the remaining issues of heredity. How did genes replicate themselves? What is the physical basis of mutation? How are genes related to illness? How are living things constructed? But before such questions could be answered definitively, it would be necessary to establish the chemical basis of heredity. Early workers such as Morgan must have had, in the words of the biochemist Erwin Chargaff, 'a perception of the ghost-like presence of chemistry in what they were doing'. But they did not have a method for elucidating life's chemistry. The solution to this key secret of existence would ultimately be found not through the agency of peas or flies, but by studying the very simplest types of creatures: bacteria and the viruses that infect them. These bugs and their parasites contained miniature instruction manuals of how to build living things. It seemed likely that the same heredity mechanisms were common to all living things, even in these much less complex organisms. If the chemical basis of a viral genetic instruction manual could be established, that of a human should eventually follow.

On the night of 17 April 1941 in London, the blitz claimed another victim: a shy and reclusive sixty-year-old bacteriologist called Fred Griffiths was killed by a German bomb. Had he survived a few years longer, he would have seen the results of a phenomenon he discovered in 1928, but chose not to investigate further, make an impact on genetics every bit as powerful as Mendel's. Griffith was interested in identifying factors that made some strains of *Streptococcus pneumoniae* – the bacteria responsible for initiating many common and often fatal diseases such as pneumonia and scarlet fever – more virulent than others. He discovered that there were many different types of pneumococci and that they differed in the composition of the sugar coat that protected them from destruction by the immune system's phagocyte cells. Bacterial strains differed in their virulence, presumably because the phagocytes found them more or less easy to destroy depending on the nature of the sugar coat.

Griffiths wondered whether he could convert a virulent strain into a non-virulent strain by incubating bacteria with an antiserum that acted on the sugar coat. He hoped that mutants lacking a coat would be unaffected by the

2. HOW TO BUILD A LIVING THING

antiserum and would survive to re-establish the population with non-virulent bugs. His attempt worked. Unlike the virulent bacteria, whose colonies had a smooth glistening appearance, the uncoated and non-virulent mutants produced rough colonies. If high doses of non-virulent pneumococci were injected into mice they had no effect: their sugarless coats made them easy prey for the phagocytes. Very occasionally, however, injection produced a lethal infection from which smooth, virulent organisms could be recovered. Griffith reasoned that the very small amounts of sugar present in the medium he had used to propagate his non-virulent strain could induce the synthesis of sugar coats in other bacteria. To prove this, he injected mice with a mixture of non-virulent living strains and heat-killed virulent organisms, neither of which produced disease when injected alone. He found that mice injected with the mixture died of infection; smooth, virulent, sugar-coated pneumococci could be isolated from their blood. During the course of the experiment the non-virulent strain had been transformed into a virulent strain by an unknown factor released from the heat-killed virulent cells. This phenomenon, a microcosm of the evolutionary process that transformed one species into another, was published by Griffith in the *Journal of Hygiene* (1928). But after a string of technical failures that prevented further analysis, Griffith abandoned the study and did not refer to it in later publications. The most important aspect of the transformation experiment – namely, that it demonstrated an inheritable change of character – was neither mentioned nor implied. Elucidation of the 'transforming principle' would have to wait for more advanced experimental techniques.

Oswald Theodore Avery, who was born in Halifax, Nova Scotia, in 1877, trained as a physician and entered general practice before deciding to give up medicine to become a bacteriologist. He joined the Rockefeller Institute in 1913, where – with the Institute's director, Rufus Cole – he planned to develop a therapeutic antiserum for pneumonia. Avery's work was disrupted by the First World War, during which time he taught respiratory medicine to army medical officers. Avery was aware of Griffith's findings but sceptical. He did not believe that one bacterial type could be transformed into another. If it was true, it would have terrible medical consequences, and might undermine his attempts at producing therapeutic antisera. But by 1930 he had replicated the transformation phenomenon and demonstrated that a non-virulent strain could be transformed into a virulent one by the introduction of a cell-free extract made from virulent cells into non-virulent cells in a test tube. But Avery became distracted by other projects and stopped working on transformation for a number of years before taking it up again in the 1940s. Through the use of enzymes that selectively destroyed certain acids – including DNA and RNA – as well as fats and proteins in the extract, he soon showed that the transformation phenomenon resulted from the transfer of genetic information from one bacterium to another. He also established that DNA was the transforming material.

2. HOW TO BUILD A LIVING THING

If the extract taken from virulent heat-treated bacteria was subjected to an enzyme that destroyed DNA before being mixed with the extract from a non-virulent strain, the mouse injected with the mixture survived. Avery and his colleagues Colin MacLeod and Maclyn McCarty published their results in the *Journal of Experimental Medicine* in 1944. In a tentative conclusion, Avery wrote that the phenomenon of transformation had been 'interpreted from a genetic point of view' and likened the inducing substance to a 'gene'. Bacteria could transfer genetic material to other bacteria through a liquid medium, and thereby transformed them, endowing them with new genetically specified characteristics. DNA was the chemical from which genes were built, and was the basis of heredity. A monumental bridge between chemistry and genetics had been constructed.

But despite the elegance of their proof, Avery's findings were seen as controversial. At the time, DNA was thought to be a biologically uninteresting and monotonous chain of four repeating chemicals which lacked the complexity necessary to encode genetic information. Proteins, which were structurally and chemically complex, seemed more obvious candidates for the genetic role. Detractors suggested that the DNA in Avery's transforming experiments was probably contaminated with protein and that the contaminating proteins were the real genetic material. DNA, in their opinion, simply provided an attachment site for the proteins from which genes were composed. This assertion, which was supported by some of the greatest chemists of the day, was hard to discount. Until 1950, no genetics textbook mentioned Avery's work.

The first step in overturning this view and restructuring the genetic landscape came from the findings of the biochemist Erwin Chargaff, who developed methods for analysing the chemical composition of DNA. At Columbia University in 1950 he showed that although the DNA in all the cells of an individual is identical, the proportions of the four nucleotide bases that constituted it vary from species to species. In Chargaff's words, the composition appeared 'to be characteristic of the species, but not of the tissue, from which they are derived'. This strongly suggested that DNA was the genetic material. Chargaff believed that combinations of the four chemicals from which DNA was composed provided sufficient complexity to form the basis of heredity. He made this clear in a 1950 issue of the journal *Experientia*. Nucleic acids from different species varied in their chemical composition, he said. He added: 'I think there will be no objection to the statement that, as far as chemical possibilities go, they could very well serve as one of the agents, or possibly as the agent, concerned with the transmission of inherited properties'. He also speculated that minute changes to nucleotides in DNA sequences could provide the physical basis of genetic mutations. But definitive evidence that the hereditary material was DNA would have to wait for the publication of the results of Alfred Hershey and

2. HOW TO BUILD A LIVING THING

Martha Chase in the *Journal of General Physiology* in 1952.

When linguists attempt to decipher ancient texts and tablets written in an unknown language, they hope they are confronted not by poetry or sophisticated religious language but by something rather more mundane and direct. The decoding of the genetic language has its parallel: without the most rudimentary examples little progress would have been achieved. Using the simplest possible versions of life – bacteria and the bacteriophage viruses that infect them (the 'eaters of bacteria') – Alfred Hershey with the help of his assistant Martha Chase, who in her spare time enjoyed sewing and complex knitting patterns, set out to determine once and for all whether the genetic material was DNA or protein. When not engaged in an experiment, Hershey could be found at almost any hour of day or night in the laboratory sitting motionless in his chair, thinking and gazing at the ceiling. He had borrowed the idea of using bacteriophages from Max Delbrück and Salvador Luria of Vanderbilt University, who had realized that bacteriophages – which had a simple protein coat enclosing a tiny piece of DNA – enabled naked genes to be studied in action. Hershey reasoned that since bacteriophages had only two components, protein and DNA, one must be genetic and the other non-genetic. As viruses usurp the metabolic machinery of the bacteria they infect, the viral genetic material must somehow enter the bacteria, with the non-genetic material that enables viruses to attach to and invade the bacteria remaining outside.

To demonstrate which was which, Hershey and Chase labelled the protein and DNA components of the bacteriophages with radioactive sulphur and phosphorus respectively. Those grown in an environment containing radioactive phosphorus produced radioactive DNA molecules; those grown in radioactive sulphur made radioactive proteins. Each of these two types of radioactive bacteriophages was then added to cultures of the bacteria *Escherichia coli*. They were given time to infect the bacteria and the cultures were then whirred in a kitchen blender, to remove any parts that remained outside the bacteria. The mixture was then put in a centrifuge to separate bacteria from any remaining intact bacteriophages or remnants of those that had infected bacteria. In cultures infected with bacteriophages grown in radioactive phosphorus, most of the radioactivity was in the bacterial pellet, suggesting that their radioactive DNA had moved from the bacteriophages into the bacteria. In those grown in radioactive sulphur, most of the radioactivity was in the liquid surrounding the bacterial pellet, suggesting that although the DNA had entered the bacteria, the protein coat had stayed outside. The protein coat was merely an instrument for injecting the viral DNA into the bacteria. DNA, therefore, was the hereditary material passed from the viruses to the bacteria. Even sceptics had to acknowledge that they had got it wrong: the identity of the genetic material had now been established beyond doubt. But how did DNA molecules encode and transmit genetic information?

2. HOW TO BUILD A LIVING THING

On 28 February 1953, eleven months after the publication of Hershey and Chase's results, the 35-year-old Francis Crick walked into the Eagle pub in Cambridge and announced in his characteristically loud voice that he and his 23-year-old colleague Jim Watson had 'found the secret of life'. It turned out that they actually had – or a key part of it anyway. They had become fascinated by the fine line 'between the living and the non-living'. When Watson, who had started his career as an ornithology student, first saw a ghostly X-ray picture of DNA presented by Maurice Wilkins at a conference in Naples, he knew immediately that it mattered deeply. Using some of the painstakingly prepared X-ray photographs taken by Rosalind Franklin and Maurice Wilkins at King's College in London, along with Erwin Chargaff's revelations about the chemical composition of DNA, Watson and Crick used plastic models to show that DNA molecules consisted of two complementary strands. These were twisted – like a spiral staircase – into a double helix. From this revelation they demonstrated that DNA could replicate itself by unzipping its two strands, using one as a template for the synthesis of the other. In 1956 the biochemist Arthur Kornberg isolated an enzyme called DNA polymerase, which was able to synthesize pieces of DNA by joining individual nucleotides together to form a polymerized chain, using one of the two unwound strands as a template. The enzyme was also able to correct copying inaccuracies occasionally introduced into the parent DNA during replication.

With the structure of DNA in place and the mechanism of its replication established, the basis of heredity was now finally clear. The information of genes was encoded in the strings of the C, T, G and A nucleotides from which each strand was made. The stability of genes resided in the fact that if one strand was damaged or mis-replicated, the other could function as a master copy to guide correction. DNA polymerase – with the assistance, as it turned out, of other 'DNA repair enzymes' – patrolled DNA sequences searching for mistakes and correcting them. The rare changes that went uncorrected due to intrinsic limitations in the efficiency of the repair enzymes became fixed into the genetic material and provided the pool of genetic variation on which natural selection could act. What also became evident was that evolution guides genetic systems to a point where they produce the optimal balance between accurate copying and inaccuracy. Accuracy was essential for the fidelity of the genetic archive, but the odd mistake in transcription is essential to provide the raw variation that enables organisms to adapt to a capricious world. The core mechanism of the evolutionary process, the source of genetic variability, had been identified.

In an instant, the erosion of mankind's innocence was complete. The time when living creatures were created exclusively by natural evolutionary means would soon come to an end. Thousands of years earlier, humans had learned to breed plants and animals. The domestication of the virtually inedible weed teosinte to make modern corn around 9000 years ago in the Balsas River basin of south-

2. HOW TO BUILD A LIVING THING

ern Mexico was mankind's first – and in some ways greatest – act of genetic engineering. Its huge ears packed with nutrients provided sustenance for whole civilizations and rapidly became a global staple. But in understanding the chemical nature and structure of the genetic material, mankind had achieved something quite different. A fundamental turning-point had been reached, from which there could be no return.

An understanding of the chemical basis of heredity does not help address questions of a more metaphysical nature. A modern answer to the reader of the first ever question-and-answer magazine, the *Athenian Gazette*, who wrote in 1691 to inquire if there are 'sexes in heaven', would not be much more helpful than the published reply, which stated confidently that in heaven all that was imperfect and accidental would be erased, so there would be no separation between male and female entities. But although it was unable to touch all areas of human existence, the rational acid of genetics would at least destroy traditional folklore notions of the nature of life. It was evident that although quite different in their organization – non-living systems had no internal representation or plan of their structure – the living and non-living parts of nature were linked by the use of the same chemicals. The secular ancient Greek accounts of how the body is made had been shown, at least in principle, to be correct. If you wanted to know how to build a living thing, it was now clear where you would have to look. The differences between giraffes, dodos, humans, zebras and pythons lay not in speculative treatises but in the digital patterns of long strings of four nucleotide chemicals, whose sequences had been honed over hundreds of millions of years as a result of chance mutations and evolution by natural selection. For centuries, philosophy had struggled to understand living things using a host of convoluted and usually untestable arguments; now, in a stroke, genetics provided a universal, rational explanation. If you wanted to know the difference between a cat and a bat, the answer lay in a comparison of their DNA sequences. All biological phenomena were unified by a magnificent principle of great beauty and simplicity.

The last part of the jigsaw necessary to establish the foundations for the future of modern genetics was the mechanism by which information stored in genetic sequences could be used to build the living things they described. How was the DNA instruction manual, written in the language of the combination of four nucleotide chemicals, translated into the protein components from which living things are constructed? This problem could be divided into two parts. First, how did the DNA code work? And second, what was the chemical mechanism that enabled this information to be read and then used to instruct the synthesis of new proteins?

Proteins are made up of sequences of twenty different amino acids. The genetic code describes how sequences of the four nucleotides in DNA determine the order in which the amino acids of newly made proteins are

2. HOW TO BUILD A LIVING THING

assembled. The uniqueness of each protein is determined by the sequence of its amino acids. Francis Crick and others showed that the genetic code was composed of triplet sequences of nucleotides, known as codons. There are 64 different codons. Each encodes either an amino acid or a 'punctuation mark' – a start or stop sign that marks the beginning or end of the gene. Since there are only twenty amino acids to encode, the code is 'degenerate', meaning that more than one codon encodes the same amino acid. Apart from a handful of minor exceptions, the code is universal and used by all known living things.

In 1961 François Jacob and Jacques Monod addressed the second part of the problem, showing that DNA does not determine protein synthesis directly. Instead, DNA is copied into a chemical intermediate known as messenger RNA (mRNA). Having made an RNA copy of a portion of the DNA, the mRNA acts as a genetic interlocutor, migrating from the nucleus to specialized cellular organelles known as ribosomes. These are microscopic machines located in the cytoplasm, which direct the process of joining up the amino acids according to the information fed into them by the mRNA, in a process known as translation. With this final piece of the jigsaw in place, the outline of the mechanism by which information flowed from DNA to proteins was complete. From the perspective of the new molecular genetic approach, natural selection could now be given a chemical basis. Mutations occurring spontaneously in genes can influence the nature of the proteins they encode if a codon, or a series of codons, is changed. The mutations themselves arise spontaneously, and although usually identified and repaired, become permanently fixed into the DNA archive on the rare occasions where they pass unnoticed. As Richard Owen had correctly predicted, science had delivered up its Newtons, Galileos and Keplers and many more besides. Together they elucidated the basis of life and evolutionary change in a way previously unimaginable. But this was just the beginning. On the shoulders of these giants, even greater things would follow.

3

Fabulous monstrosity

On Wednesday 11 February 1863, the *New York Observer* published an account of 'the event of the century, if not unparalleled in history'. This was the wedding of 'the most extraordinary and interesting MAN IN MINIATURE in the known world', Charles Sherwood Stratton, a twenty-five-inch high midget who weighed only fifteen pounds, to the thirty-two inch Miss Lavinia Warren, an 'astonishing specimen of minute humanity'. At a time when England was in the thrall of the imminent marriage of the Prince of Wales, New York had conjured up its own state occasion. Stratton, also known as General Tom Thumb, was already a celebrity and immensely wealthy, having toured the world with the Victorian freak show impresario Phineas Taylor Barnum, charming all with his entertaining anecdotes, his imitations in 'FULL MILITARY COSTUME' of Napoleon Bonaparte, and his performances of a variety of songs and dances, especially the hornpipe and the polka. He had been received at Buckingham Palace by Queen Victoria no fewer than three times, and at the Queen Dowager Adelaide's palace had been presented with a miniature gold watch and chain.

By 11 o'clock, an hour before the 'Fairy Wedding' was

due to commence, the streets around Grace Church were overflowing with expectant crowds. All stagecoaches and vehicles were diverted from the route by official order. The church was packed with New York's great and good, the congregation resembling 'brilliant tulips ablaze in the mid-day sun' and the organist playing a selection of marches including a theme from *Tannhäuser*. There was a grand entrance, described by the *New York Herald* as a 'carnival of crinoline, the apotheosis of purple and fine linen'; and the tiny bride and groom positioned themselves on a specially constructed platform to the right of the pulpit. The Rev. Mr Willey of Bridgeport, Connecticut, conducted the ceremony. They were then conveyed into a carriage and, after passing through a throng of animated well-wishers who nearly blocked Broadway, were taken to the Metropolitan Hotel, where they greeted their guests from the top of a grand piano to a serenade by the New York Excelsior Band. The wedding gifts displayed in an adjoining room included a miniature silver tea set, a tiny billiard table with matching balls and cues, an intricate silver sewing-machine and a doll-sized suite of ebony and gold furniture.

Phineas Taylor Barnum had pulled off yet another stunt, once again exploiting the public's fascination with nature's explorations on the theme of human form, and had further increased his great wealth. With the help of General Tom Thumb he had captured and formalized the Victorian interest in human exotica and brilliantly packaged it. Before long, such oddities as the ectrodactylous

3. FABULOUS MONSTROSITY

Lobster Boy, Fred Wilson – whose hands were split down the middle (his middle fingers and metacarpal bones were missing) – were paraded as a matter of course in shop fronts, museums, music halls, travelling fairs and circuses. Barnum's exhibits included such self-made marvels as the tattooed ladies Mary Brooks and Nora Hildebrandt and such run-of-the-mill variants as the Bearded Lady, Madame Devere, and the Living Skeleton, J.D. Avery. But there were more bizarre specimens, such as the Frog Boy, Avery Childs, whose Ehlers-Danlos syndrome enabled him to adopt extraordinary positions; and Jo Jo, the Russian Dog-Faced Boy, who suffered from hypertrichosis and was covered from head to foot in thick hair. Such wondrous examples of unexpected anatomy and profound deviations of human form in some cases resulted from disruptions to the processes of embryological development and included Charles Tripp, the Armless Wonder; the conjoint twins Millie and Christine, headlined as the 'Two-Headed Girl'; the phocomelia or 'seal limbs' of Eli Bowen, the Legless Acrobat; and, most famously, the Elephant Man, James Merrick.

Such deviations from expected anatomy as the Two-headed Girl and the Armless Wonder are intriguing and tantalizing. They promise an insight into the deeper logic underlying the process of embryonic development; their very existence indicates the immense power and potential for variability inherent in biological generative mechanisms. But examples like these are both rare and random. In cases like that of the Russian Dog-Faced Boy, where unusual traits run in families and alteration of form is a result of genetic modification, a study of DNA from affected individuals can help elucidate the causes of their unusual appearance. But in those that arise spontaneously as a result of chance mutations and abnormal developmental processes, genetic studies are less helpful. And many of the mutations that produce dramatic alterations to body structures result in spontaneous abortions, making examples all the rarer. For these reasons, the study of naturally occurring human oddities has not greatly helped to advance our understanding of how we are made. If humans were like peas, we could cheerfully mutate human embryos and use the resultant bestiary of broken form to investigate normal generative processes. But fortunately – though inconveniently – we are not like peas, and so must look for assistance elsewhere.

The forces that generate living creatures from the amorphous collection of cells that constitutes the early embryo are now known to be encoded in DNA. But the quest for this elusive form-giving agency has a long and convoluted history. The quest begins in antiquity. The notion that the assembly of a systematic catalogue of nature's varieties of animal form could help address fundamental issues of human existence first emerged in the *Physiologus* (literally, 'the whole sermon of nature'), a seminal Greek text of uncertain provenance thought to have been written in Alexandria in the third to fourth century AD. Unlike earlier texts attributed to Aristotle and Pliny the Elder, the *Physiologus* attempted not just to

3. FABULOUS MONSTROSITY

describe the natural world but to interpret it, within the allegorical context of Christian doctrine. You might call the *Physiologus* the first self-help manual. While conducting its large survey, it looked to the natural world for principles to correct the folly of human existence, artfully combining the natural history of animals with moral principles derived from them. Unfortunately, no original manuscripts survive, the earliest remaining versions being Latin translations. Over the centuries, the text changed beyond recognition. Prose mutated into poetry, new creatures of questionable credibility – such as gryphons – were introduced, and the number of chapters increased unchecked as popular treatises of the day were cut and pasted into the original. In later editions the text was embellished with illuminations and lavish illustrations, of which the Aberdeen bestiary is a fine example.

The 1492 version of the *Physiologus*, written by Bishop Theobald, Abbot of Monte Cassino from 1022 to 1035, and first printed in Cologne, is a good example of the way animals were used for moral instruction. Christ was indicated by the nature of the lion and the Devil by that of the fox. The leopard's spots and the tiger's stripes signified men stained with vice and hypocrisy. The medieval author Gregory, for example, wrote that 'for truly every hypocrite is a tiger, in that while he derives a pure colour from pretence, he is striped, as it were with the intermediate blackness of vicious habits'. Bestiary monographs did little to further the understanding of the principles that generated the form and behaviour of the creatures contained within their pages. Indeed, the Aristotelian doctrine that menstrual blood was sparked into spontaneous life by semen persisted unchallenged well into the seventeenth century.

The early bestiaries focused on natural forms of animals, but in the sixteenth and seventeenth century it became fashionable to include 'human prodigies' or 'monsters' in the assorted collection of natural and mythical creatures. One of the first of the new generation of bestiaries was *Des monstres et prodiges* ('On monsters and prodigies'), published in 1573 by Ambroise Paré, a surgeon to the French royal family. The inclusion of human marvels was initially meant merely to record their existence and to provide a further target for moral sermons; deviations from normal human form were taken as a sign of divine displeasure and a symbol of man's corruption. But Paré went further than his contemporaries. Although 'the wrath of God' remained a possible explanation for human deformities, he considered the possibility that terrible things might ensue if people had sex with animals or while menstruating. He perpetuated the commonly held belief that the mere sight of deformity by a pregnant woman could induce a similar misfortune in her developing child, and speculated that narrow wombs and inappropriate amounts of seminal body fluids might be a cause of deformities. He was wrong on all counts, but the agency responsible for the generation of life had been brought into the domain of the rational. The shaky but nevertheless discernable foundations of an enterprise that

3. FABULOUS MONSTROSITY

would eventually lead to the science of genetics had been assembled.

The first person to appreciate the scientific value of natural variations was the free-thinking Francis Bacon, who entered Trinity College, Cambridge, at the age of twelve, and went on to become Lord Chancellor. His *Novum organum*, first published in 1620, was a turning-point in natural science: it repudiated the validity of received authority, ancient learning, speculation and religious doctrine. Bacon wrote that knowledge of the natural world was replete with 'vulgar notions, scarcely beneath the surface'; scholars and the clergy were anxious and obstructive lest there should be 'something discovered in the investigation of nature to overthrow, or at least shake religion, particularly among the unlearned'. He demanded 'that the entire work of the understanding be commenced afresh', guided by observation and logical deduction from empirically derived core principles; he dismissed the Aristotelian notion that final causes (teleology) guided the patterns in nature. With his *res non verba* – things not words – as their motto, the 'new philosophers' who followed him revisited the empirical and atomistic strands of ancient Greek thought attributed to the 'mechanical philosophy' of Epicurus, Democritus and Lucretius, a philosophy that had largely been sidelined in favour of Aristotelian logic. The body was now a *machina carnis* – a machine of the flesh – and living forms could be viewed mechanistically, as ingenious contraptions, devices, pieces of machinery constructed from skilfully fashioned components.

Bacon distinguished between three broad categories of nature. These were 'freedom of nature' (normal nature including natural variations), 'errors of nature' (abnormal variations of nature) and 'bonds of nature' (artificial nature) – or what Bacon called 'the Mechanical and the Experimental Art'. Bacon's key revelation was that 'errors of nature' could be used to infer general constructional principles that might in the future be used to produce artificial or mechanical life. This thought was revolutionary, being both an extraordinary biological insight and the first modern statement of the natural science agenda, which would eventually result in the discovery of the gene and the real possibility of redesigning life from first principles. In order to realize this agenda, Bacon proposed that one 'must make a collection or particular natural history of all the monsters and prodigious products of nature, of every novelty, rarity or abnormality'. He did not suggest a specific method for extracting generative principles from the collection, but knew that 'once a nature has been observed in its variations, and the reason for it has been made clear, it will be an easy matter to bring that nature by art to the point it reached by chance'. The drama of this statement lay not only in its anticipation of the science of genetics, but in its heretical assertion that form originated as a consequence of chance events rather than from the design of a divine Creator.

Bacon's rationalist approach became incorporated into

3. FABULOUS MONSTROSITY

the great crisis of European thought known as the Enlightenment, or Age of Reason: an organic period of profound intellectual change that transformed mankind from a premodern society dominated by superstition, folklore, witchcraft, unchallenged ancient scholastic opinion and Christian doctrine – with its judgment, fear, punishment and eternal torment – into an essentially modern and liberated scientific enterprise. The changes then occurring in every area of life – from fashion to the food fads of Lord Byron – represented a shift of ideology in which the religious doctrines that had permeated every area of natural study and had hampered empirical investigation were replaced by a new open-minded approach that resulted in an emancipated and secularized view of life. The long-held notion that the blood of a hanged man had curative properties was now discounted as a 'vulgar error'. God became a 'distant causer of causes', and it was no longer acceptable to attribute phenomena to supernatural influences. The Enlightenment led to many 'modern' concerns and preoccupations: the existential anxieties arising from the erosion of belief that led Alexander Pope to complain of 'this long disease my life'; the notion of moderation propounded by the Scottish physician George Cheyne, who, at thirty-two stone, was so fat that he needed a servant to walk behind him carrying a stool on which to recover every few paces; the cult of the individual; and the championing of the mind as an engine of liberation.

Bacon's subversive agenda encouraged others to explore the causal elements underlying the development of human anatomy. The first significant step was the work of the physician William Harvey published in 1651: his insights into the origin of form – at that time referred to as the problem of generation – were derived from experimental observations that enabled him to challenge a specific Aristotelian doctrine, namely, that menstrual blood was sparked into life by semen. Harvey had the good luck to marry the daughter of one of Elizabeth I's physicians; the royal connection was to be of great importance to his scientific career. In 1618 he was appointed court physician to James I, a job that continued under the patronage of Charles I until 1649, when the king was executed by Oliver Cromwell. Six years before his death in 1657, Harvey published his great work, *Exercitationes de generatione animalium* ('Essays on the generation of animals'), in which he invented the science of embryology, the precursor of genetics three centuries later. This relied heavily on his years of royal patronage, which had given him unlimited access to pregnant does taken from the king's deer parks. Careful dissection of their wombs revealed no evidence of fauns coagulating out of menstrual blood in the manner predicted by Aristotle. He did, however, observe fluid-filled sacs. They were absent in deer that had not mated; Harvey speculated that they were equivalent to the externally laid eggs of hens. He was in fact looking at early embryos rather than eggs, but his conclusion – *ex ovo, omnia* – that everything comes from the egg, although at the time speculation, turned out to be correct, and of singular importance. Harvey could not himself answer the question of how form emerged from an amorphous egg, but succeeded in defining the problem. And he gave it a name: epigenesis.

Despite these findings, subsequent anatomists ignored Harvey's notion of epigenesis and the idea that form was generated *de novo* from a formless state during a process of intrauterine development. Instead they reverted to the ancient scholastic doctrine of preformation. This stated that the reproductive organs of females were populated either by the 'essential parts' or 'patterns' of the adult structures, or by miniature and perfectly formed primordial human forms that later became known as homunculi (from the word homunculus, 'little man' in Latin). From this perspective, new life originated when a somnolent homunculus was triggered by fertilization to grow and increase in size. This was in spite of the work of early microscopists such as the Italian Marcello Malpighi, who went on to become personal physician to Pope Innocent XII. Malpighi's *Dissertatio epistolica de formatione pulli in ovo* ('Dissertation on the formation of the chicken in the egg'), published in 1673, described the embryological development of chicks in great detail. He did not, however, accept that epigenesis was inevitable, since he could not rule out the possibility that a representation of the final form was present at a sub-microscopic scale in the unfertilized egg. The Dutch microscopist Jan Swammerdam went further, speculating that if a microscope of sufficient power could be constructed, human forms of diminutive size would be visible in the ovaries long before fertilization.

These ideas were formalized by the French philosopher Nicolas Malebranche, whose *De la recherche de la vérité* ('On the search for truth'), published in 1674, became the rallying cry for preformation, arguing that the reproductive organs were stocked with a vast collection of miniature homunculi, all created by God at the beginning of time and merely waiting to be activated. A later variant of preformation was the notion of spermism, which – in opposition to ovism, which argued for the primacy of the egg – held that the homunculi resided in the heads of sperm stored in the testis. The belief in preformation, which was to inform embryology for almost a century, even found its way into the literary canon. In *The Life and Opinions of TRISTRAM SHANDY Gentleman*, published as a series of volumes in the 1760s, the ailing satirical writer Laurence Sterne imagined what it must be like to be a homunculus. He gives a graphic description of Tristram's conception following his father's ejaculation, which 'scattered and dispersed the animal spirits, whose business it was to have escorted and gone hand in hand with the HOMUNCULUS, and conducted him safe to the place destined for his reception'.

Harvey's theory of epigenesis was eventually resuscitated, largely as the result of a chance discovery made by a young Swiss amateur naturalist, Abraham Trembley, who was tutor to the sons of Count William Bentinck, a Dutch nobleman, at his estate near the Hague. To pass the time, Trembley had taken to collecting samples of water from local ponds and ditches, storing them in jam jars and observing them under his microscope. He came across a small creature a few millimetres long which he referred to as a polyp. Although green, suggesting it was a plant, it moved from one location to another. Uncertain whether his polyps – later renamed hydras after the mythical Hydra whose nine serpent-like heads spontaneously regrew when severed – were plants or animals, Trembley decided to cut them in two. His journal of 25 November 1740 reads as follows. 'The first operation I performed on the polyps was to cut them transversely', which he did while holding 'a little water in the shallow of my right hand' before severing the body with scissors. He was amazed to see each fragment regrow into an entirely new polyp, in a phenomenon that came to be known as 'regeneration'. Trembley published his findings in 1744. They appeared in a beautifully illustrated volume entitled *Mémoires pour servir à l'histoire d'un genre de polypes d'eau douce* ('Memoirs concerning the natural history of a type of freshwater polyp'). It was difficult, if not impossible, to explain the phenomenon of regeneration from the perspective of preformation. And the declaration, in 1839, by the German zoologists Theodore Schwann and Matthias Schleiden that all living creatures were comprised of cells suggested that there might be cells specialized for the task of generation. But the definitive blow to preformation came with Ernst von Baer's announcement in 1828 that the human ovum, or egg, was a single cell with no space for a preformed homunculus.

Trembley's work on the regeneration phenomenon was

followed in 1759 by the publication of an equally important volume, *Theoria generationis* ('Theory of Generation'), written by the German physician Caspar Friedrich Wolff. Wolff described the essential germ layers – endoderm, mesoderm and ectoderm – from which an embryo is made. He could not explain the nature of the *vis essentialis* or 'life force' that underlay the form-giving process of epigenesis, but restated epigenesis's central thesis, asserting 'that the organs of the body have not always existed, but have been formed successively – no matter how this formation has been brought about'. The link between Wolff's *vis essentialis* and modern genetics was forged by the unlikely agency of the mid-eighteenth to early nineteenth-century *Naturphilosophie* movement, trumpeted by the scientist and author Johann Wolfgang von Goethe, who originated the term 'morphology' and was a proponent of comparative anatomy. Goethe's combination of German romanticism – in which the unconscious spirit (*Geist*) of 'lower' creatures struggled to attain the conscious state of man – and scientific realism, focused exclusively on empirically verifiable phenomena, outlined the modern biological agenda which would eventually lead to the genetic paradigm.

The *Naturphilosophie* movement attributed the *vis essentialis* to a poorly defined metaphysical agency, but it also insisted on a careful and systematic microscopic examination of developing embryos in order to throw light on the epigenetic process. This spawned the science of embryology, the study of embryos as they develop from

3. FABULOUS MONSTROSITY

amorphous single-celled entities into complex multicellular structures. *Naturphilosophie*'s focus on comparative anatomy, which argued for an underlying unity in biological form, along with its attempts at ordering creatures into series according to their complexity and its quasi-metaphysical notion that species strive towards perfection, together made it an important precursor of the evolutionary theories of Jean-Baptiste de Lamarck and Charles Darwin. Darwin's great achievement was to strip the mysticism from Lamarck's evolutionary theory, outlined in his *Philosophie zoologique* (1809), according to which the *besoin*, or need for improvement, led simple creatures to be transformed over time into more complex ones. Darwin also initially argued that modifications to structures that had occurred during an individual's lifetime could be inherited.

Darwin's theory proposed that the process of inheritance was imperfect, resulting in random minor modifications to structures that provided a repertoire of variation from which the environment could select the best-adapted examples. As the contours of the environment changed over time, buckling the 'adaptive landscape' in which species evolved, so the survival value of any given variant would change according to the nature of its environment, both physical and social. In Darwin's model, evolution was an inevitable consequence of the incremental change brought about by the accretion of multiple small variations, generating new species over time. Although freaks or 'prodigies' of nature might help illuminate the agency of development, these 'hopeful monsters' played no role in the continuous graded variations that were the raw material for everyday evolutionary processes.

While a student in Cambridge in the late 1820s, Darwin, like many other Victorian gentlemen, was caught up in the beetle craze sweeping the nation. Beetle-collecting seemed to be his only real passion. Many of his spare moments were spent wandering across local meadows, fens and ditches with his cousin William Darwin Fox and Fox's dog, Little Fan. Essential items for their hunting expeditions included a sweeping-net and a copy of the beetling bible, William Kirby's *An Introduction to Entomology*, which Darwin carried under his arm. Darwin's great triumph as a student was the discovery of a rough-bodied black beetle with red antennae: it turned out to be a rare German species, *Melasis flabellicorni*, which had been spotted in England only twice before. His find was no doubt a huge blow to his rival Charles ('Beetles') Babington. The apparent infinity of beetle species encouraged the accumulation of huge bestiaries and was an apparent testament to the diversity of God's Creation. In Darwin's later life, details of the gradated variations within and between closely related species such as beetles and finches would form the cornerstone of the incremental gradualism that underpinned his formulation of evolutionary theory. He spent much of his life amassing examples of variations that fitted into his incremental

3. FABULOUS MONSTROSITY

scheme, where minor variants were perpetuated and augmented on the basis of their adaptive value.

Darwin's collections of variants were assembled with an eye for the tiniest variations. The bizarre bestiary of the anti-Darwinist William Bateson was quite different. Described by his headmaster at Rugby as a 'vague and aimless boy', at university in Cambridge Bateson became absorbed with zoology and morphology, studying among other things the embryology and evolutionary origins of the obscure worm *Balanoglossus*. Bateson felt uncomfortable with Darwin's concept of gradual change, since what struck him was not the continuity between species but the 'extraordinary discontinuity of variation' and the vast gaps between different designs. Darwin had provided a very simple explanation for this dilemma: the intermediate forms of species that once filled the gaps between apparent discontinuities had become extinct. But Bateson remained sceptical, convinced that nature's discontinuities must arise from mechanisms intrinsic to organisms and occurring independently of any superimposed environmental sorting. This led him to state that 'variation, in fact, is evolution'. Bateson was happy to postpone the quest for the causes of variation, which was in his judgment premature, deciding instead to undertake an extensive forensic examination of its minutiae. The results of his exhaustive cataloguing enterprise were published twelve years after Darwin died, in *Materials for the Study of Variation* (1894), when Bateson was thirty-three. The volume, whose thesis that species evolved by a series of discontinuous jumps was controversial, was not a success, and the proposed second volume was never written. But the contents of the bestiary Bateson assembled proved to be of great importance in helping to unravel the processes by which normal and abnormal form are generated in developing embryos.

The collection Bateson assembled was the scientific equivalent of a Victorian freak show, containing hundreds of bizarre deviations from normal anatomy. Bateson divided the examples in his compendium of animal and human abnormalities into three broad categories. 'Substantive' variations were of the classic continuous Darwinian kind, whereas 'meristic' variations involved a change in the number of body parts. 'Homeotic' variations involved a reorganization of body parts, with one appendage appearing in place of another. Meristic and homeotic variations were the basis of his theory of discontinuous evolution. Discussing one example of meristic variation, a mutant longicorn *Prionidae* beetle with twelve-jointed antennae instead of the usual eleven, Bateson argued that the proposition that the new joint had been acquired gradually would lead to an 'endless absurdity'. A half-made joint would clearly be useless, demonstrating that the transition from eleven to twelve joints must have been discontinuous. The most dramatic examples in his bestiary, however, were the result of homeotic variations. One such was Case No. 76, a specimen of the honeybee *Bombus variabilis* 'taken beside the hedge of a park in Munich, having the left antenna partially developed as a foot'. It was implausible that classic Darwinian evolution could have selected for a creature that had a foot growing from its head. To Bateson such examples were clear evidence for the discontinuous way in which bodies were modified. The key to understanding evolution and the modification of body structures lay not in the mediocrity of mundane variations, but in fabulous monstrosity.

Bateson's volume was one of the last great biological texts that relied exclusively on observation. As the century turned, the experimental assault on the *vis essentialis* began in earnest, culminating in Watson and Crick's elucidation of the three-dimensional structure of DNA. But though Bateson was shunned by the Darwinians for being disloyal, and criticized for not suggesting a mechanism to explain discontinuous change, his focus on the importance of homeotic modifications, irrespective of their actual role in evolution, was visionary. The genetic phenomenon underlying these bizarre homeotic alterations to body form was to be immensely rewarding for the scientist Edward Lewis, whose studies of the phenomenon would bring him a Nobel Prize in 1995.

Homeotic mutations in flies – such as the *bithorax* mutation which resulted in an extra pair of fully-formed wings producing a four-winged fly; *bithoraxoid*, which produced an eight-legged beast rather than the usual six-legged variety; and *antennapedia*, in which legs sprouted from the head in place of antennae – were already well known to Lewis when he started his work in the 1930s.

Through mating studies using flies, he was able to show that the genes responsible for these dramatic alterations to the fly body plan were clustered together to form what he called the homeotic gene complex, or HOM-C. This cluster of master control genes appeared to program the development of flies, assigning identity to regions of developing tissue and determining whether they developed into an eye, a leg, a wing and so on. Lewis speculated that the HOM-C gene cluster had arisen by the duplication of an ancestral homeotic gene which had evolved around 500 million years ago.

In the early 1980s, at Stanford University, David Hogness and Welcome Bender were able to isolate the so-called *hox* genes responsible for such major alterations to the body plan. By 1983, Walter Gehring and William McGinnis had shown that *hox* genes shared a common region of DNA sequence, called the homeobox. The homeobox enables *hox* genes to perform their key function, which is to switch other genes on and off. While Gehring was isolating *hox* genes in flies and worms, his colleague Eddy De Robertis, working on the same floor, wondered whether Gehring's success in flies might be emulated in vertebrates. He was working on the common frog *Xenopus laevis*, so decided to try an experiment that seemed crazy. He looked for *hox* genes in frogs. There was absolutely no reason to believe that vertebrates would use the same genes to build themselves as invertebrates. His colleagues were sceptical and a number of students refused to participate.

3. FABULOUS MONSTROSITY

But De Robertis was soon celebrating: frogs did have *hox* genes. The first genes likely to play a significant role in vertebrate development had been discovered. *Hox* genes were subsequently also identified in mice and humans. The astonishing finding resulted in what was called the 'hox paradox'. If humans share many of their master developmental regulatory genes with creatures as diverse as flies and sea urchins, then what are the distinguishing factors? It seemed implausible that homeotic mutations were responsible for the evolution of new forms: it was more likely that evolution had performed subtle 'rewirings' of the networks of genes controlling development. These changes would result in the same set of genes producing different patterns of activity, creating different chemical 'morphogenetic' fields in the developing organism. Cells exposed to these different chemical landscapes would, much like miniature computers, calculate the differences and behave differently in each instance. These changes in regulation were likely to be responsible for producing the array of animal architectures we are familiar with.

The fact that the same *hox* genes were used to generate the body plans of creatures as different as elephants and ants was itself extraordinary. But *hox* genes exert their influence late in development, programming the final plan of organisms which already have their front and back defined and which have been divided into segments. Studies of *hox* genes would not help unravel the mysteries of the very earliest moments of development, when form emerges from an undifferentiated ball of embryonic cells. Insights into these processes awaited the arrival of another bestiary. The zoo of twisted and deformed mutant flies assembled by Eric Wieschaus and Christiane Nüsslein-Volhard, which provided a more graphic representation of living hell than anything imagined by Hieronymus Bosch, was produced artificially, unlike Bateson's collection of naturally occurring oddities. Not content for nature to produce its own slow and erratic mutations, Wieschaus and Nüsslein-Volhard exposed flies to chemical mutagens which – like a shotgun fired at pieces of paper – tore into the DNA sequences of the *Drosophila melanogaster* fruit flies, introducing a host of random mutations. The result was a collection of over a million embryos taken from the mutated parents of around 40,000 fly families. These were studied one by one, and the mutant genes responsible for the resulting developmental abnormalities hunted down.

This study, published in *Nature* in 1980, showed that around 5,000 of the fly's 20,000 genes were important in development, around 140 being essential. The genes could be divided into three categories: gap genes, which acted at the very earliest stages to lay down the organism's basic template; pair-rule genes, which divided the body into segments; and segment-polarity genes, which affected specific structures within the segments. Despite the power of this approach in revealing the underlying logic of pattern formation in a complex organism, a fly was not a human – or even a fish, a frog or a mouse. Flies were invertebrates. Although many of the fly genes involved in development were also found in vertebrates, it seemed likely that there was a whole universe of other developmental genes involved in constructing the organs and anatomy present in vertebrates but absent in flies. The problem was that it would not be easy to create a bestiary of mutant frogs or mice. The beauty of flies was that they were tiny, reproduced rapidly, and generated large numbers of offspring which could easily be stored in jam jars.

The breakthrough came through the use of a diminutive, rapidly breeding, transiently transparent but otherwise unremarkable species of zebrafish called *Danio rerio*. Nüsslein-Volhard reasoned that the technique that had been so successful with flies – randomly disabling genes and observing the consequences – could equally well be applied here, and the faulty fish offspring could be used to piece together the blueprint of a vertebrate. The creatures' tiny size made the proposition practical. Nüsslein-Volhard immediately set about constructing a huge aquarium or '*fischhaus*', containing some 6000 tanks, in her laboratory at Tübingen. After adding mutagenic chemicals to the tanks of males, she ended up with the mutant embryo zoo she had hoped for. A survey of this universe of embryonic wreckage, containing over 1.2 million specimens, uncovered a great richness of developmental errors affecting every stage.

These included mutant genes affecting very early development, such as *half-baked* and *bozozok* (Japanese for 'arrogant youth on motorcycles'), which impaired the process of gastrulation – when the featureless bowl of cells in the early embryo divides into a tube-like body that separates into segments. Another mutant, *knypek* (Polish for 'short'), produced pancake-like embryos that resembled trilobites. Other mutants had abnormal behaviour: *techno trousers* twitched uncontrollably; *backstroke* swam in continuous circles. Others, such as *unplugged*, were virtually paralysed. But the elucidation of the complete network of interacting genes that computes complex form remains an outstanding challenge, even for a simple fish. One of the problems has been the time it takes to track down the damaged genes. In 2002, however, Nancy Hopkins and her colleagues announced a new method for tracking down the underlying mutant genes. This involved using viruses, tagged for easy identification, to disable zebrafish genes. The tag on the virus could then be used to pinpoint the underlying mutation. The enterprise that started long ago in Alexandria with the *Physiologus* now promises to deliver the genetic programs and mechanisms that will eventually enable living things to be built from first principles.

4

Smart genes

Professor John Henry Pepper was a flamboyant showman and celebrity of the Victorian age. His books, such as *A Boy's Playbook of Science* (1860), were instant bestsellers. Families flocked to London's Royal Polytechnic Institution, where for a sixpence they could marvel at steam engines, automata and fully operational diving bells. But the greatest attraction of all was the Institution's chief lecturer, Professor Pepper himself, with his conjuring tricks and amazing experimental demonstrations. In the 1880s Pepper toured Australia, Canada and the United States, bringing his exhibition to a global audience. He helped shape the public imagination at a critical moment in the history of science, adapting his repertoire from the experiments of famous scientists of the day such as Michael Faraday. But Pepper's renditions were always fantastical, making mundane phenomena spectacular and realising them on a gigantic scale. The induction coil added to the performance in 1869 produced a spark almost a metre long. But his most famous trick was what came to be known as 'Pepper's ghost'. Pepper gives an account of this in his *The True History of the Ghost and All About Metempsychosis* (1890). The performance opened with his adopted son walking onstage, bowing and addressing the audience:

> 'Ladies and gentlemen, I am sorry to inform you that something has detained Professor Pepper.' At that point a voice could be heard. 'Stop, stop: I am here!' And appearing out of nothing and without the aid of trap doors or descent by the help of copper wires, the author appeared in the midst [of the stage] and bowed his acknowledgment for the hearty greeting kindly given him by his audience. The metamorphoses then proceeded. Oranges were changed into pots of marmalade and a chest of tea was converted into a steaming teapot, sugar, milk, cups of tea and handed by the attendant to the ladies in the reserved seats only. The ghost of Banquo in Macbeth and the ghost of Hamlet followed. This was then followed with the curious change of a deserted piano into one at which played and sang a living member of the fair sex, attended by a gentleman in faultless black coat and white tie, who turned over her music; and this Part I wound up with the change of the gentleman into a lady, who walked down to the footlights, sang a song, and then vanished into 'thin air'.

Following its debut in a Christmas performance of Charles Dickens's the *Haunted Man* in 1862 and many other appearances, Pepper's ghost was eventually co-opted by fairgrounds, including Disneyland's Haunted House ride, and the horror film classics of directors such as Alfred Hitchcock. The realistic ghosts conjured up by Pepper in his shows amazed Victorians, many of whom were obsessed with mediums and spiritualism. But Pepper

4. SMART GENES

had other intentions. Using his 'ghost-generating' apparatus – consisting of a set of screens, special lighting and a huge sheet of concealed glass carefully positioned at a forty-five degree angle – he aimed to shatter the public's delusions. By showing that the apparently impossible could be achieved through simple means, he demonstrated the ease with which the human mind could be deceived into attributing the results of simple tricks to metaphysical forces. Pepper's ghost illustrated that when the underlying rational principles have been understood, the mystery disappears and the impossible becomes effortlessly routine. His show was a metaphor for scientific progress, embodying the optimism of a world newly galvanized by the fruits of scientific experimentation and the triumph of rationalism over the supernatural. If Pepper's ghosts had such a simple explanation, might our own nature yield to such analysis?

Of all the world's phenomena, the human mind is without doubt the most complex and intriguing. It is not surprising that it has always been difficult to conceive of the mind as a simple property or embellishment of an already complex machine. The very existence of something so remarkable seems to demand an explanation transcending anything formulated in the language of science. In the West, the Christian agenda of death, judgment, heaven and hell appeared a sufficient explanation for body and soul, accounting not just for the present but the future, well beyond our brief sojourn on Earth. Magic, superstition and custom were further adjuncts to the narrative of existence. In Dante's fourteenth-century *Divine Comedy*, humanity was an integral but merely dependent lever in God's great Creation. Surveyed in the finest detail by this omniscient consciousness, even the smallest deviation was subject to punishment. It was not until after the Enlightenment that man could become *faber suae fortunae*, the author of his own destiny, and acquire the independent consciousness alluded to by the sixteenth-century essayist Michel de Montaigne in his phrase *arrière boutique toute nostre* – a room behind the shop all of our own. The outline of the modern notion of self – the significance of the individual, the preoccupation with our festering imperfections, the existential crisis issuing from the loss of religious certainty – had been firmly established. Writers such as Jean-Jacques Rousseau, in his *Confessions*, were now free to bare their souls not just to priests but also to a voracious and ever-expanding public. Rousseau's invention of the device of personal revelation – in his case of vice, sexual deviation, masochism and cruelty to children – reverberates today in the pages of tabloid newspapers, reality television and the cult of the celebrity.

The French 'mechanical' philosopher René Descartes cleverly avoided offending the clergy by asserting that the mind had a rational embodiment, while simultaneously allowing for its metaphysical aspects. But it was the discovery at the turn of the nineteenth century of genes and gene theory that promised to resolve the mysteries of living machines once and for all. If the *vis essentialis* was

4. SMART GENES

embodied in genes, it was clear that the secret to life must lie in an organism's entire gene set, or genome. The genome is the apparatus – like Pepper's ghost-generating machine – that builds individuals of a species. If it were possible to define the contents of genomes, the parts list for the assembly and operation of organisms would be revealed. In the case of man, the inventory would include specifications for the components that make brains and minds, raising the possibility that philosophical issues of the type addressed by Descartes might yield to an empirical dissection. The independent invention by Walter Gilbert and Fred Sanger in the mid-1970s of techniques for determining the genetic sequences of organisms provided the means for elucidating genomic descriptions. The early methods were slow and clumsy, but in 1977 Sanger nevertheless sequenced the entire genome of the tiny *phiX174* bacteriophage. This proved the general feasibility of the genome-sequencing concept. The difference between the genome of a tiny virus and that of a human was simply one of scale. But the sequencing of *phiX174* had been slow and painstaking. A technological breakthrough would be necessary if the genomes of complex organisms were to be attempted.

Although now routine, the concept of genome sequencing on an industrial scale – and, in particular, the idea of sequencing the human genome – was highly controversial when it was first publicly mooted in the mid 1980s. Robert Sinsheimer, at that time the Chancellor of the University of California, Santa Cruz, was one of the first people to appreciate its importance. Influenced by the plans of Santa Cruz astronomers to build the world's biggest telescope, he decided that biology needed a project of similar magnitude. But there was a fundamental problem. At an estimated three billion nucleotides long (in other words, containing three billion chemical 'letters'), the human genome was 20,000 times larger than the Epstein-Barr virus, which at that time was the largest genome to have been sequenced. In 1985 Sinsheimer gathered together some of the world leaders in the field to explore the feasibility of a human genome sequencing project. These included John Sulston and Bart Barrell from Cambridge University and Walter Gilbert from Harvard. All agreed that the idea was bold and exciting, but that there were several problems.

The first of these was that the type of 'big biology' required for the project's success ran contrary to the cottage industry style and hypothesis-driven spirit of traditional biochemistry. Critics argued that sequencing the human genome would be nothing more than a grand 'fishing exercise', requiring armies of mindless 'worker bees' – more like factory employees than scientists. Sydney Brenner, an early proponent of genome sequencing at Cambridge, joked only half-heartedly that genome sequencing could be subcontracted to prisoners as a type of punishment. Another objection was the project's price tag, estimated to be in the region of $3 billion. Funds for science were already drying up. David Botstein, a geneticist at the Massachusetts Institute of Technology, stated that the project's threat to drain resources 'endangers all of us, especially the young researchers'. Then there was the objection that as much as 98% of the human genome contained 'junk' – DNA sequences that lacked protein-coding genes. What was the point of sequencing the whole thing? Sydney Brenner suggested that one option was to sequence the compact genome of the pufferfish, which was an eighth of the size of the human genome, containing the same number of genes but including far less 'junk'. The most significant obstacle, however, was the apparent technological implausibility of the proposition. Many believed the barriers to be so significant that the project would be unrealizable.

Thanks to the advocacy of Charles DeLisi, a cancer biologist at the US Department of Energy (DOE), and other influential supporters such as James Watson, a Human Genome Initiative was eventually established at the DOE in 1986. The DOE's interest in genome sequencing was streamlined with their studies of radiation and human health. But there was concern that a biological project of this scale had been mooted by an agency other than the National Institute of Health (NIH), which hitherto had presided over all major US public healthcare projects. The turning point came with the favourable report of the National Research Council (NRC) panel chaired by Bruce Alberts, published in February 1988. This not only endorsed the project but redefined it, recommending that full-scale sequencing of the human genome should be delayed until sequencing became faster and cheaper. In the meantime, analysis could begin on the genomes of simpler organisms – such as the bacteria *Escherichia coli*, baker's yeast *Saccharomyces cerevisiae*, roundworm *Caenorhabditis elegans*, fruit fly *Drosophila melanogaster* and mouse *Mus musculus*. These would help annotate the human genome by enabling inter-species comparisons. By contrasting the genome of a fly with a worm, for example, it would be possible to identify both shared genes and those unique genes that give worms their 'wormness' and flies their 'flyness'. The genomes of living things would no longer be black boxes with uncharted depths of genetic complexity. Genes would be counted, archived and subjected to detailed forensic analysis. For the time being human geneticists would content themselves with making genomic maps – signposts that flag up genomic regions and help locate genes associated with human diseases. This strategic repackaging of the human genome project resulted in Congress allocating it $28 million in 1988.

NIH rapidly established an office for genome research and recruited James Watson to head it. It wrested control of the project from the DOE and set about assembling a consortium of international collaborators. Watson proved an excellent advocate, summarizing the project's goal as being 'to find out what being human is'. But he was adamant that the other key objective was to produce a gene sequencing machine equivalent in power and virtuosity to the particle accelerators of atomic physicists. Everything ran smoothly until June 1991, when J. Craig

Venter, who ran a large sequencing laboratory for NIH, announced an audacious plan. Venter's idea was to skim the cream off the human genome, focusing on gene-encoding sequences and ignoring the rest. In collaboration with his colleague Mark Adams, he had devised a method that enabled genes to be identified at lightning speed. In this, small regions of sequence were obtained from the tail-most end of each messenger RNA molecule and used as a tag for the whole gene.

These small bits of sequence, known as expressed sequence tags (ESTs), represented only the genes that were switched on. If all of them were sequenced, the majority of the protein-encoding genes would be captured. Venter argued that this approach was 'a bargain in comparison to the genome project', enabling around 90% of human genes to be identified for a fraction of the cost of the entire genome sequence. Watson was incandescent: he complained that 'virtually any monkey' could adopt Venter's strategy and told Congress that he was 'horrified'. Most objectionable to Watson was the fact that Venter, with NIH funding and their blessing, planned to file patents on each of the new EST sequences – despite their small size and in many cases having no idea of their function. Watson's protestations to his NIH boss, Bernadine Healy, were ignored; shortly afterwards, he was released from his duties. In 1991 Venter also left the NIH, accepting an offer of $70 million from private investors to set up his own not-for-profit EST sequencing initiative, the Institute for Genomic Research (TIGR).

Following Watson's departure, the helm of the NIH National Human Genome Research Institute in Bethesda, Maryland, was taken up by Francis Collins, who rode to work on a motorbike and played guitar in a rock band. He was a trained physician, and his priority was to identify genes associated with diseases by pushing forward with the mouse and human genome maps. These proved invaluable in helping to locate disease genes and functioned as a stopgap while the complete human genome sequence was being completed. The NIH received a significant boost when the Wellcome Trust established a new sequencing laboratory in Hinxton near Cambridge. But the technological breakthrough that everyone had hoped for never materialized and the human genome sequencing continued at a snail's pace.

In July 1995, just as things seemed to be faltering, the sequencing community received the news that Craig Venter, Hamilton Smith, Rob Fleischmann and Claire Fraser had determined the genome sequence of a free-living organism, the prokaryotic (lacking a nucleus) bacteria *Haemophilus influenzae*. This had been achieved in record time using a method known as whole-genome shotgun sequencing, in which the entire genome was shredded into small pieces. These were then sequenced and reassembled using computer software that recognized overlapping ends. This approach was dramatically different from the conservative and far slower method adopted by the public consortium, which worked with smaller, well-defined pieces of DNA. But there were many who did not believe the shotgun technique would work: the NIH had refused Venter funding for it at an early stage in its development.

In 1996 the NIH made its own dramatic announcement: the completion of the genome sequence of the yeast *Saccharomyces cerevisiae*. This was especially significant, since, like the complex eukaryotic cells of humans and other animals, yeasts have a nucleus and distinct cellular organelles. Encouraged by this advance, John Sulston, who had made major inroads into the sequence of the roundworm *Caenorhabditis elegans*, tried to persuade Collins to let him tackle the human genome, arguing that if they settled for marginally less accurate data the project could be completed more rapidly. Collins would have none of it, insisting that the definitive 'book of life' had to be accurate. But on 9 May 1998 everything changed. Venter announced he had teamed up with the private Perkin-Elmer Corporation. With the help of 300 of their newly automated sequencing machines and one of the world's fastest supercomputers, they started up a private company: Celera Genomics, based in Rockville, Maryland. They planned to sequence the human genome, junk and all, within a remarkable three years. This was not only significantly faster than the public consortium, but at only $300 million, considerably cheaper. The task would be accomplished using the shotgun technique tested so successfully on *Haemophilus*. The shock of the prospect of being beaten by a renegade private company galvanized the NIH to alter both their priorities and strategy. In

September 1998 Collins announced that a 'rough draft' of the human genome, covering 90% of its sequence, would be available by 2001. This was the same timeframe as Venter's, and would be achieved using the shotgun method. The race between the public consortium and private company was on.

Collins aimed to undermine the commercial value of Venter's data by ensuring that the NIH sequences were instantaneously available to the public. This would significantly reduce the number of sequences Celera would be able to patent, since once in the public domain the 'novelty' criterion necessary for patent filing would not be upheld. On 11 December 1998, only months after Venter's announcement, John Sulston published the sequence of the flatworm *Caenorhabditis elegans*, the first genome of a multicellular animal. In the meantime, in order to cut his teeth on a larger genome and further test the shotgun method, Venter started work on the fruit fly genome. This was done in collaboration with a publicly funded team led by Gerald Rubin at Berkeley in San Francisco and polished off with great speed. Its publication in March 2000 demonstrated beyond doubt that the shotgun method could deliver. The competition reached boiling point as the race neared its final stages.

Although Ari Patrinos of the DOE had managed to broker some encouraging behind-the-scenes negotiations between the two parties, and had mooted a proposal for a collaborative effort, things went awry when the Wellcome Trust leaked a letter to the press in March 2000, in which

Collins informed Venter that there were irreconcilable philosophical differences between their approaches. The main problem related to the way sequence information would be made available. After a meeting in February 1996, the public consortium had established a code of conduct known as the 'Bermuda Rules', which declared that human genome sequence information should be given away for free. Patenting was discouraged and data release would be rapid and unconditional, the key doctrine being that the human genome belonged to everybody. Although Celera had agreed to give not-for-profit academic researchers free access to its data, it would make commercial institutions pay for access to its databases and would negotiate specific deals to secure rights in their use. Celera information would be deposited not in the public GenBank sequence database but on the secure company website. These arrangements were a core part of the Celera business plan and formed the basis on which it had raised its funds from private investors. Even if Celera had wanted to revise its position, shareholders would never have allowed it.

In a brief rapprochement in June 2000, Venter and Collins, chaperoned by Patrinos, shook hands with President Clinton at the White House and pledged to support one another. This prompted *Time* magazine to claim 'the race is over'; but the situation soon deteriorated. Plans to collaborate were rapidly forgotten, though the parties did agree to co-ordinate their publications. Two independent human genome sequences were finally published in 2001, with the public consortium's data appearing in *Nature* on 15 February 2001 and Celera's in *Science* on 16 February 2001. The race was over: both sides were victors. Richard Gibbs, who headed the sequencing project at the Baylor College of Medicine in Houston, described his experience of viewing the completed human genome sequence as 'the same feeling you must get when you are on a satellite ... looking down at Earth'. In future all biological experimental results would have to be explained with reference to this. The definition of the basic inventory of components needed to build a human meant that the possibility of additional layers of genetic components could be disregarded. This was the entire gene kit. The human genome sequence was rapidly followed by that of the mouse *Mus musculus* in December 2002 and rat *Rattus norvegicus* in April 2004. A host of other organisms were subsequently completed, including a wide range of microorganisms, the pufferfish *Fugu rubripes*, sea squirt *Ciona intestinalis*, tiny weed *Arabidopsis thaliana* and the malaria-carrying mosquito *Anopheles gambiae*.

In the topsy-turvy world of genetics, the unexpected can often be anticipated. But little could have prepared anyone for the staggeringly small number of genes found in the human genome. Rather than the predicted 100,000, which would have matched intuitive notions of our complexity, the initial count revealed as few as 32,000. Francis Collins described this as 'a bit of an assault on our sensibility'. Things got worse in June 2003, when the tally was downgraded to a paltry 24,500 genes. Our entire

4. SMART GENES

genome might be twenty-five times as big as that of a fly, but we possess fewer than twice the number of genes needed to make a fly. More disturbingly, we have fewer than the 25,498 genes needed to make the weed *Arabidopsis thaliana* and only around one-third more than the 18,424 used by a worm.

The notion that gene number alone could explain biological complexity had been exploded. It was replaced by what was known as the N value paradox, where N refers to the numbers of genes in a genome. If the number of genes in organisms is so small, and so similar between different species, where does the information responsible for the differences between species come from? A measure of biological complexity other than gene number was needed. Part of the problem related to the difficulties associated with the definition of a gene and the uncertainty surrounding the apparently geneless or 'dark matter' regions of genomes. Gene-prediction computer programs are biased towards detecting active genes: those that are only rarely or slightly expressed are easily overlooked. And although the classic definition of a gene is a region of DNA encoding a protein, some regions of DNA are now known to encode small regulatory RNA molecules known as microRNAs (miRNAs). Many of these – despite being made of RNA rather than protein – have regulatory functions. The inclusion of miRNA-encoding genes would doubtless change the relative tallies significantly.

As might be expected, the evolution of organismal

complexity employed an array of tricks to generate advanced computational capabilities. The increase, or in some cases decrease, in gene numbers is only one example of a whole armoury of such devices. (Its importance should not be underestimated, since the presence or absence of even a single gene can have a huge impact.) And having the same number of genes, or even broadly speaking the same genes, does not mean that they are identical. A point mutation can have a dramatic impact on gene function. The size of genes and the complexity of the protein structures they encode also vary considerably between species. But it soon became clear that the human genome sequence was not the definitive description of an organism's information. This was not to say that gene numbers do not give a ballpark indication of an organism's complexity. Transitions of complexity, for example, from simple prokaryotic organisms to complex eukaryotic organisms, from a unicellular state to a multicellular state, or from invertebrates to vertebrates, are all accompanied by incremental increases in gene numbers. But within each group the number of genes in different species is broadly similar. So whereas the gene numbers in bacteria vary between 500 and 8,000 and overlap with those of single-celled eukaryotes, more complex multicellular eukaryotes such as worms and flies have gene numbers in the range of 12,000 to 14,000. The genomes of vertebrates such as humans, fish, mice and rats contain around 25,000 genes.

If gene numbers alone cannot account for the range of

4. SMART GENES

complexity in a group of related creatures, and if the differences in gene numbers between simple and complex organisms are less striking than anticipated, then what are the differentiating factors? The main difference between the genes of 'higher' organisms, such as vertebrates, and those of 'lower' organisms, such as bacteria, is that they are 'smarter'. This does not imply that their genes are aware of what they are doing, but simply that each gene is more complex, as is its repertoire of behaviour. If we want to identify the factors responsible for the graded levels of complexity within a group of organisms like vertebrates, it is to the relative 'smartness' of the genes that we must defer. As genes become smarter, the organisms they build and operate become more complex. Genes do not act in isolation, but work together in circuits that co-operate to generate higher-level functions. The more flexible the circuit, the more sophisticated the behaviour it performs. The difference in relative 'smartness' between gene circuits is similar to the differences between a pocket calculator and a desktop computer, or a personal computer and a supercomputer. As with artificial calculating devices, the complexity of the genetic components in biological systems corresponds with the sophistication of the computations or 'algorithms' they are able to perform.

The smartness of genes relates to the complexity of the mechanisms that regulate their expression and the synthesis, nature and activity of the proteins they encode. This smartness has many manifestations. Generally speaking, the smarter a gene, the greater its capacity to respond to nuances in the chemical environment surrounding cells and to make different proteins – a trick achieved by modifying either the transcribed mRNA or the protein itself. Eric H. Davidson, at the Californian Institute of Technology, has described these complex regulatory mechanisms as the 'brain' of the smart gene. The importance of such regulatory sequences was discovered in the 1920s by Alfred H. Sturtevant and Hermann Muller, who described a new type of mutation in fruit flies. The 'position effect' mutants they studied resulted not from changes to the physical structure of the genes, but to their altered positions within the chromosome. The discovery that genes could be affected by changes to their positions on chromosomes undermined the concept of a gene as a simple particulate entity and led Leslie Dunn in 1937 to state that the gene was showing 'signs of disappearing in a cloud of position effects'. One of the advantages of assembling genome sequences from multiple species is that the non-coding regions (the 'junk' regions) can be compared, in order to trace the 'footprints' of the regulatory sequences that have functional significance. There is likely to be far more lurking in the non-coding 'junk' than was originally anticipated.

The most basic of the mechanisms that regulate gene expression are the control elements which determine when, where and at what stage in an organism's development a gene is switched on. Various sequences are responsible for these control functions. The main type, known as 'promoters', are situated close to the beginning

4. SMART GENES

of genes. 'Enhancers', on the other hand, may be located at great distances from the coding sequence they influence. Genes may be associated with multiple enhancers, each of which conveys a discrete element of 'intelligence', influencing where, when and the extent to which the gene under its control is activated. These enhancer modules, each with a different control function, represent additional tiers of gene processing complexity. Enhancers exert their effects on gene activity by interacting with the gene's promoter.

The unsophisticated genes of the simplest organisms are turned on and off like simple switches. The smarter genes of higher organisms, however, have evolved complex promoter control panels that sense many aspects of the cell's environment, integrating external biochemical signals in order to determine whether the gene should be off, on, or on to some extent. Other elements of the promoter determine the stage of the organism's life cycle in which the gene is activated. Each element in it works like a logic gate in an electronic circuit, requiring certain input conditions to be satisfied before activating the gene they control. In other words, the more complex promoters are in effect genetic microcomputers. The 'program' or software that runs these tiny DNA computers is directly encoded in the regulatory elements or 'modules' from which each promoter is made. Eric Davidson and Chiou-Hwa Yuh of the California Institute of Technology have shown that the promoter controlling the activity of the *Endo 16* gene in the sea urchin *Strongylocentrotus purpuratus* contains more than thirty different binding sites for more than thirteen different transcription factors (proteins that promote gene transcription). The modules that make up the *Endo 16* promoter work as the gene's on-off switches and volume controls, and each is associated with its own suite of transcription factors.

If we assume that each gene can be either on or off (a huge oversimplification, since the degree to which a gene can be on is continuous rather than digital), then a genome with N genes theoretically encodes 2^N expression states. Humans – with their 24,000 or so genes – could in principle produce $2^{24,000}$ different gene expression patterns of on and off states. A creature like a fly, on the other hand, having only 14,000 genes, has the potential to produce only $2^{14,000}$ states. Each state should be imagined as being associated with a unique property or behaviour. So the greater the number of possible states, the higher the potential sophistication of the underlying genome. The difference between the complexity of a fly and a human can be explained not only by the 10,000 more genes found in humans, but in terms of the number of different gene expression states each genome is capable of producing. The difference ($2^{24,000}/2^{14,000}=2^{10,000}$) is a huge number, larger than the number of elementary particles in the known universe. This means that a relatively small variation in the number of genes between two species has the potential to generate a tremendous difference in biological complexity. This is because the repertoire of gene expression states available to a genome constitutes the executive part of the 'genetic computer'.

The dramatic changes that can occur when genes are activated in inappropriate anatomical locations were demonstrated in a remarkable experiment performed by Walter Gehring at the University of Basel in Switzerland. The discovery came by chance in 1993, when one of his students, Rebecca Quiring, was screening a library of fly genes for transcription factors. Transcription factors are specialized genes encoding proteins that bind promoters; in so doing, they influence the level of gene activity. One of the candidates fished out by Quiring was shown by Uwe Walldorf, a post-doctoral student in the laboratory, to map to a region of the fly chromosome known to encode a gene called *eyeless*. When *eyeless* was mutated, affected flies had eyes that were either deformed or absent altogether. The discovery that *eyeless* encoded a transcription factor reminded Gehring of a phenomenon he had studied thirty years earlier while a graduate student at Zurich University. In flies and other insects, adult structures such as wings, eyes and legs arise from embryonic tissues called imaginal disks. If these were removed from developing flies they could be grown in flasks, usually retaining their identity. But occasionally their identity shifted: a wing disk might form an eye. The mechanism for this was never determined, but it had been assumed that master genes controlled developmental programs and that if one of these were accidentally switched on it could cause one program to override another.

Following up on his hunch that *eyeless* might be just such a gene, two of Gehring's students, Patrick Callaerts and Georg Halder, engineered flies so as to switch *eyeless* on in disks destined to become wings, legs and antennae. When the flies started to hatch, some of them had eyes on their wings. Others had eyes on the ends of their antennae which looked 'like little crab eyes' on stalks. This extraordinary discovery showed that changes to the timing and location of the expression of a single transcription factor gene could result in dramatic anatomical changes. Given the importance of transcription factors in interacting with promoters to regulate the expression of genes, one might expect that the number of genes encoding transcription factors would scale with the complexity of the genome under consideration. And that is exactly what is found. The genome of an anatomically simple worm encodes only 500 transcription factor genes, flies have 700, and the human genome more than 2000. The surprising fact that the genomes of plants encode around 1500 transcription factors indicates that the smartness of animal genes does not reside exclusively in the complexity of their promoters and protein transcription factors. Nor do these figures include miRNA transcription-like factors.

The startling discovery made by Victor Ambros in 1993 that miRNAs could influence gene expression in the worm *Caenorhabditis elegans* added a new dimension of complexity to the processes by which the information of genes is transferred to proteins. It turns out that both plants and animals utilize miRNA control factors. Javier

4. SMART GENES

Palatnik showed that an miRNA molecule called *JAW* influences the shapes of plants. If the gene encoding *JAW* is mutated, it causes the leaves of *Arabidopsis* to curl up rather than lie flat. The relative complexity of miRNA processing in different species remains to be seen, but it is predicted that around 1% of mammalian genes encode miRNAs in addition to mRNAs. This means that the number of miRNAs in the human genome is roughly equivalent to the number of transcription factors, a clear indication of their importance. The discovery of miRNAs has meant that the notion that genes encode only proteins has had to be revised. miRNAs have functions every bit as important as proteins. Despite being found in virtually every eukaryotic species studied, there is no evidence that miRNAs are present in simple organisms such as yeast and bacteria. Unlike protein transcription factors, which bind to promoters, miRNAs recognize target sequences within mRNA molecules and cleave them. This has the effect of inhibiting the translation of the mRNA messages. Strictly speaking, miRNA molecules are 'post-transcriptional control factors' rather than transcription factors. In April 2004 Soraya Yekta, at the Whitehead Institute of Biomedical Research in Cambridge, Massachusetts, showed that miRNAs can influence the expression of *hox* genes in mouse cells, indicating that – as with *JAW* in plants – miRNAs play a critical role in the development of vertebrates. But the smartness of genes does not end with transcription factors, miRNAs, enhancers and the logic of promoters. Many other tiers of complexity serve to generate even more information from the core set of genes within a given genome.

In 1963 the American paediatrician John Bruce Beckwith and his German collaborator Hans Rudolf Wiedemann reported an unusual fetal overgrowth syndrome, known as the Beckwith-Wiedemann Syndrome (BWS). Affected babies are gigantic, with features including an enlarged liver and spleen, an enlarged tongue, ear lobe creases and asymmetrical growth of various body regions. Around 10% of these children develop Wilms tumour, a form of kidney cancer. The good news is that BWS is one of the few disorders that improve with age; Wilms tumour is treatable if caught early, and the majority of BWS children are perfectly normal once they are adults. BWS is remarkable for not being caused by a DNA mutation, but by a process that provides an insight into another type of genetic regulation found only in mammals.

Every new individual inherits two functionally equivalent copies of each gene, one version from the mother and the other from the father. A phenomenon known as genomic imprinting is, in a select group of genes, able to silence one of these two identical genes without changing the DNA sequence itself. Imprinted genes are 'haploid', having only one active copy rather than being 'diploid', with two active copies. Interestingly, the imprinted genes are derived from either the mother or the father but never from both. The expression of imprinted genes is therefore determined by their parental origin. The imprinted insulin-like growth factor-2 *Igf2* gene responsible for BWS is, for example, always expressed from the paternal version; the maternal version is silenced.

The presence of imprints that switch expression off in select groups of genes means that although the maternal and paternal genomes are roughly equivalent from the point of view of their DNA sequences, they are functionally different. A DNA sequence description of the human genome is therefore incomplete unless we know which genes have been imprinted and so inactivated. Imprinting itself is caused by the selective chemical modification of cytosine nucleotide residues in the DNA of the affected gene, in a reversible process called 'methylation'. Changes in gene expression brought about by methylation are known as 'epigenetic' modifications; the complete description of which genes in a genome are methylated is called its 'epigenotype'. Wolf Reik and his colleagues in Cambridge demonstrated in 1997 that the features of BWS result from the dosage effect of having two active copies of the *Igf2* rather than the normal single active paternal copy with an accompanying imprinted (methylated) and inactivated maternal copy. The way in which the methylation of genes inhibits their expression is itself intriguing.

The DNA of chromosomes is not free-floating: it is wrapped around groups of eight specialized proteins called 'histones', to produce a histone octet known as the 'nucleosome' (or 'chromatin'). The nucleosome, once thought to function as an inert scaffolding to package

DNA, is now known to have a much more dynamic role, influencing the access of other proteins, such as transcription factors, to the enclosed DNA. It functions much like the chaperones who used to accompany Victorian ladies on their excursions, determining whether the DNA is allowed to have access to visiting proteins or not. If the histones are configured so as to obscure the genes from such proteins, the gene promoter is blissfully oblivious to transcription factors and remains switched off. Whether or not the histones allow the gene to have access to other proteins is influenced by whether the histone has itself been chemically modified. If the histone is methylated it does not permit gene expression, and it also represses the gene if it is stripped of its normal acetylation (another type of chemical modification). Methylated genes attract a protein called MeCP2, which exists as a complex with a deacetylase enzyme. This strips off acetyl groups from the histones and silences the gene. Histones are subject to other chemical modifications, such as phosphorylation, and it is thought that the complex combination of acetylation, methylation and phosphorylation defines a histone code that fine-tunes gene expression. DNA sequence information alone is therefore very incomplete. The epigenotype and histone code are every bit as essential in defining an organism's genetic information.

The smartness of imprinting is that it enables organisms to recruit into their developmental programs genes which would normally have adverse effects if the proteins they encode were at their normal high, double dose concentrations. By silencing one copy of the gene, the concentration of the protein they encode is halved, making their use in developmental circuits a practical proposition. The adverse effects of higher doses of the proteins encoded by imprinted genes are seen in the various diseases that result when imprinted genes are stripped of their methylation to produce 'methylation mutants'. Imprinting errors are just as significant in development as DNA sequence mutations. Imprinting is a mechanism that is specific to mammals and may have been important in the evolution of complex structures such as the brain. But methylation of both gene copies is a widely used strategy in both animals and plants, though not in some species such as yeast and the fruit fly *Drosophila melanogaster*. Each cell in the human body has a full set of the 24,500 or so genes in the human genome, but each cell type – a nerve cell, say, or a skin cell – gets its particular identity by expressing only a defined subset of the genes it contains. Methylation is used to switch off genes unnecessary to the operation of the sub-program that generates and sustains particular cell types. Methylation was originally thought to be involved only in long-term silencing of gene expression, but in October 2003 Yi Sun and his colleagues at the University of California at Los Angeles showed that methylation is used in the genes of neurones to adjust their responses to brain activity.

Promoters, enhancers, transcription factors, miRNAs, imprinting, histone modifications and methylation add multiple tiers of complexity to the basic information represented in the gene inventory of a species, and operate at the level of gene expression. But there are still further mechanisms that increase the smartness of genes. The genes of eukaryotic organisms are not contiguous but are stored as segments known as 'exons' separated by intervening bits of DNA known as 'introns'. mRNA molecules are made by copying both introns and exons and then splicing the exons together. Some species are able to cut and paste the exonic segments of their genes in different ways, by a process called alternative splicing. These different splice forms result in a single gene being able to make multiple proteins – in much the same way as a film can be multiply edited, producing the standard version and the 'director's cut'. Around 40% of genes in humans produce alternative mRNA molecules, which in turn encode different versions of the protein. In simpler eukaryotic organisms, such as yeast, a much lower percentage of genes do this, and to a lesser extent.

The number of alternative molecules produced by a given alternatively spliced gene can vary from two to over ten. The gene *Troponin T*, expressed in muscle, is unusual in generating more than eighty different splice forms; the relative abundance of each of the different forms varies according to the type of muscle and the prevailing physiological conditions such as growth and exercise. Errors of alternative splicing have been shown to underlie various diseases. The mRNA molecules transcribed from genes can also be altered by a process called RNA editing. Shin Kwak and his colleagues at the University of Tokyo demonstrated in February 2002 that the fatal paralytic disease amyotrophic lateral sclerosis is probably a result of a disorder of RNA editing. The efficiency and stability of mRNA molecules can also be affected by a process called 'alternative polyadenylation', in which the sequence following the tail end of genes can be substituted for different versions.

At the level of proteins, a whole additional layer of complexity is introduced on top of all these genetic mechanisms. Proteins can be processed and modified in a host of different ways, or whisked away to different cellular locations. These 'post-translational modifications' range from the common, such as phosphorylation, ubiquitination, methylation, acetylation, glycosylation (in which sugars are added to the protein), to the more obscure, such as sulfation, hydroxylation, epimerization, sumoylation, neddylation and glutathionylation. Mutations that affect post-translational modifications of proteins have been linked to diseases. Proteins also bind to one another and to other proteins, sometimes self-assembling into composite structures. Remarkably, it is not only genes that can be spliced. James Yang and his colleagues at NIH showed in January 2004 that, like mRNA molecules, some proteins can also be directly cut and spliced to generate new forms. These tricks and modifications, coupled with self-assembly and self-organizational processes, generate a dynamic combinatorial library of properties that vastly increases the complexity of the protein repertoire.

The term 'proteome' has been used to describe the complete catalogue of all the different proteins and modified versions of proteins that a genome can generate. As organisms become more complex, their genes become smarter; and so do their proteins. Perhaps the greatest trick of all, however, is the development of the 'delegated' complexity of the vertebrate immune system and brains. From a small repertoire of genetic instructions, these sub-systems generate immense capacities for information; in the case of the human brain this is almost unlimited. In the human brain, the storage of information in a biological substrate has been transcended, with information being stored outside genes in cultural artefacts such as music, art, books and computers.

5

Reprogramming life

As nineteenth-century London grew in size, so did the amount of rubbish produced by its inhabitants. Much of this consisted of the residuum of fires, including white ash, cinders and fragments of unconsumed coal. A lucrative dust removal business was rapidly established, with rival suppliers paying large sums to secure contracts. Henry Mayhew estimated in the 1850s that there were around 1800 dustmen or 'dusties' in London. The dust was taken away in carts to dust-yards in suburbs near canals or rivers, such as Bermondsey and Shadwell. Sifters would then separate finer particles from the coarse ones, removing any objects mixed in with the dust. Fine dust was sold for use in fertilizers; the coarser material was used in the firing of bricks. Many things were found when sifting through the dust, including 'oyster shells, old bricks, old boots and shoes, old tin kettles, jewellery, old rags and bones', money, garbage and other junk. The dust was piled into huge conical heaps, having 'much the appearance of a volcanic mountain'. The sifters, generally the wives or mistresses of the dustmen, were paid a shilling a day and allowed to keep half the value of anything they found. They often pocketed their finds, but faced immediate discharge if caught. The women, who were usually middle-aged, wore 'black bonnets crushed and battered like those of fish-women', coarse dirty cotton gowns and strong leather aprons upon which they repeatedly struck their sieves. On visiting a dust-yard, Mayhew noted the cocks and hens 'scratching and cackling amongst the heaps' and the numerous pigs that seemed 'to find great delight in rooting incessantly about the garbage and offal collected from the houses and markets'.

Like a Bermondsey dust heap, the genomes of living things are replete with junk. Around 99% of the human genome consists of DNA sequences that do not encode proteins and which were initially thought to lack a function. But this is not any old junk: in the words of the Harvard biologist Walter Gilbert, 'you can find lovely things in it'. The geneticist Sydney Brenner described junk as the 'stuff you're glad you saved because it still has value'. The idea that junk was uninteresting arose mainly because the kinds of simple organisms, particularly bacteria, studied intensively in the 1950s and 1960s are wall-to-wall with genes and contain little junk. This meant that geneticists focused on protein-encoding sequences; the rest was dismissed as detritus. When the Human Genome Project was first mooted many argued that sequencing junk was a waste of time and resources. Fortunately, the whole lot – junk and all – was eventually sequenced. Genomes do not just consist of tiny islands of genes in vast seas of irrelevant junk. Each chromosome is best thought of as a complex information organelle, containing sophisticated control and maintenance mechanisms. The genome itself functions as a dynamic biochemical machine. Although the information to make the components of living things resides in the sequences of genes, a huge amount of information lies outside genes, scattered throughout the 'junk'. This non-genic machinery is responsible not just for the maintenance and regulation of genetic processes but for their reprogramming.

The junk found in genomes is not homogeneous. It is more like the junk found in dust heaps, and divides into discrete categories. Around 17% of the human genome, for example, consists of junk sequences known as retrotransposons. Retrotransposons are uniquely mobile pieces of DNA, able to make copies of themselves and integrate at different sites in the genome. They are examples of a class of mobile genetic elements known as transposons, first identified by Barbara McClintock in the 1940s. Most of the transposons in eukaryotic genomes are of the retrotransposon type. These 'jumping' or transposable DNA segments have altered our understanding of how genomes are structured and modified. Given their importance and abundance, it is extraordinary that their discovery took so long. Transposable elements are of central importance in restructuring genomes in every organism studied so far, from plants and bacteria to yeast, flies, worms, mice and humans. They constitute a large fraction and in some cases a majority of the DNA in many species of plants and animals. Incredibly, more than 50% of the maize genome is made from transposable elements. The accumulation of retrotransposons between genes explains the difference in genomic size between maize and its smaller relatives, indicating that in some species transposons play a significant role in expanding genomes.

Bateson and others established Mendel's laws as the central paradigm of genetics. Central to these was the notion that genes are discrete, immobile elements, neatly arranged in a linear fashion along chromosomes, each with a fixed and invariant position. Within this framework, the main sources of genomic variation were recombination between chromosomes (in sexual organisms), in which matching regions in a pair broke off and exchanged positions, and the slow and incremental accumulation of point mutations. But this classic view of genetics was inconsistent with the results of an experiment conducted on the maize plant *Zea mays* in summer 1944 by Barbara McClintock. Her discovery, totally at odds with Mendelian dogma, resulted from her interest in the broken ends of chromosomes. She had been working with a strain of maize that had repeated breakages on chromosome 9, leading to the loss of a chromosomal fragment during development. This consistent breakage at the same dissociation site (Ds) contrasted with the more familiar pattern of random breakage. The key observation was that maize plants with breakages at their Ds had multicoloured kernels.

After closely studying the behaviour of the Ds using markers, McClintock was able to show that any given Ds was able to transpose itself, or jump, from one position to another. She soon linked the mobility of this region to the mutations resulting in multicoloured patterns of kernel colouration. McClintock hypothesized that transposable elements like Ds were always present in the maize genome but that under normal circumstances they were suppressed or relatively silent. When genomes were subjected to extreme stress such as chromosomal damage, however, their activity was unleashed, resulting in multiple mutations occurring rapidly. The way in which the Ds caused the increased mutation rate was not known, but it was clear that gene expression was being modified in some way. It was later demonstrated that the mutations were caused by the Ds moving to a new downstream location on chromosome 9. This was accompanied by a large inverted duplication of the intervening chromosomal segment. These changes disrupted genes that synthesized the pigment, and gave the kernels their unusual appearance.

McClintock had discovered that DNA can jump from one position to another, and that genomes are programmed to respond to stress or damage. Emergency responses of this kind presumably evolved to enable genomes to rapidly alter gene expression and reorganize themselves when their survival was threatened. This can lead to profound and irreversible changes over very short timescales. It is likely that they were of great significance in reprogramming genomes and in producing the changes in gene expression and chromosomal organization necessary for the evolution of new species. In cases where transposons were relatively silent, such changes would be rare. But when transposons were activated, in times of stress or damage, significant reprogramming could occur over small numbers of generations. The activity of individual retrotransposons is controlled both by epigenetic factors and by the composition of their sequences.

When McClintock first presented her discovery of transposons (which she termed 'controlling elements') at a meeting in 1951 at Cold Spring Harbor, the audience responded with a mixture of 'puzzlement' and 'hostility'. In her Nobel Banquet speech of 10 December 1983, she said that her work had been 'ignored, dismissed, or aroused frustration'. For many years she was rarely invited to give a seminar, let alone a lecture. At the time when she was doing her most important work she was ignored to such an extent that she chose not to publish her results. But when transposons were eventually independently discovered in bacteria, yeast and other organisms there could be no denying that she had been right all along. The full significance of her monumental discovery became apparent; McClintock became famous and was awarded a Nobel Prize.

Retrotransposons replicate by making a piece of RNA and converting it into DNA using an enzyme called reverse transcriptase. Selfish retrotransposon 'parasites' have accumulated in mammalian genomes over millions of years, making copies of themselves and jumping into new locations. There are two main types of retrotransposons: long interspersed elements (LINEs), which are up to 6,000 nucleotides long, and short interspersed elements (SINEs), which are around 300 nucleotides long. If the integration site of a newly created retrotransposon happens to be a gene, the gene may be disabled or its function altered. Newly integrated retrotransposons can also delete large segments of DNA from their insertion sites, carry adjacent sequences along with them, activate previously dormant genes and cause sequences to become inverted. Such deletions and modifications generate some of the raw material of evolutionary change and are an important source of genomic plasticity. Retrotransposons may also alter the sites at which homologous regions of chromosome pairs recombine with one another, or provide new sites of similarity that encourage recombination. Retrotransposon activity varies considerably between species, however. Mercifully, retrotransposon activity quietened down in humans around 40 million years ago and today fewer than 100 actively jump around human genomes. But in mice things are different, with around 3,000 retrotransposon sequences still on the move. It has been estimated that retrotransposons are responsible for around 10% of the mutations that produce noticeable changes in mice characteristics.

As their insertion sites are random, many of the mutations caused by newly created 'young' retrotransposons are harmful. In order to protect themselves from these potentially damaging effects, genomes have evolved mechanisms for excluding retrotransposons from essential regions. Areas housing *hox* genes, for example, are kept almost entirely free of retrotransposons. This means that they tend to accumulate in places where they do not get in the way and cause damage. Older retrotransposons also have an effect in reprogramming genomes: they are able to alter the expression of neighbouring genes in a graded manner, providing yet another level of control. Jeffrey Han and his colleagues at the Johns Hopkins University School of Medicine in Baltimore, Maryland, showed in May 2004 that retrotransposons can function as 'molecular rheostats', finely tuning gene activity by setting the parameters within which they operate. This is achieved by a mechanism dependent on the unusual ratios of nucleotide bases in transposons which bog down the transcription machinery and prematurely truncate mRNA molecules transcribed from the gene. The significance of retrotransposons in gene expression was shown by Han's demonstration that 79% of human genes have retrotransposon sequences located within their non-coding 'intronic' regions. Genes associated with intronic retrotransposons produce much less mRNA; large genes often contain many scraps of different retrotransposons.

The idea that junk contained some kind of message was tested by the physicist Eugene Stanley of Boston University. Using statistical techniques borrowed from linguistics, he showed in December 1994 that junk sequences were non-random and had features of ordinary language. This hinted 'that something interesting was lurking in the junk'. Retrotransposons are good examples of bits of junk DNA that have a definable function, but they are by no means the only ones. The genome is full of very short sequences known as microsatellites in which one to six nucleotides are multiply repeated. These monotonous sequences epitomize the banality of junk. But in August 1993 Theodore Krontiris at Tufts University School of Medicine in Boston showed that mutations in microsatellite sequences were implicated in up to 10% of all cases of breast, colorectal and bladder cancer. The virtuosity of junk DNA does not end there. In November 1993 Lisa Sandell in Seattle removed the repetitive DNA sequences at the ends of chromosomes, and found that the chromosomes disintegrated. This suggests that repetitive sequences help maintain the integrity of the genome. Others have confirmed that 'gene deserts' within the human genome harbour sequences with important regulatory functions. Working at Berkeley in California, Marcelo Nobrega showed in October 2003 that these regions contain long-range enhancers which influence the expression of genes located far away.

Although intrinsically more difficult than assigning functions to genes, decoding junk is an important task and is likely to provide insights into how evolution remodelled the genomes of complex organisms. The initial strategy for identifying functional non-coding sequences involved foraging in the DNA junkyard and looking for the effects of mutations. But the availability of genome sequences from a collection of different species has resulted in a new approach. This involves identifying the footprints of regulatory sequences by comparing non-coding regions from multiple species. As evolution relentlessly tinkers with genome sequences, testing the results by natural selection, functional junk should, be conserved across species. The closer any two species, the more non-coding regulatory sites they should share. But if species are too close, many pieces of DNA will be conserved by chance. These approaches have demonstrated the dramatic way in which genomes have been restructured. In a number of cases the sequences of protein-encoding genes are duplicated. Duplications were important in evolution, as one master copy was retained to perform the core function while the other was free to accumulate mutations and develop new functions. The sequences of parent and daughter genes inevitably diverge as the daughter is modified. In some cases the daughter gene is inactivated. Genes which have lost the ability to synthesize a protein-encoding mRNA or an miRNA are known as pseudogenes. Sequencing studies have shown that duplicate genes are abundant in most genomes and significant portions of genomes are repeated. Wen-Hsiung Li and colleagues at the University of Chicago showed in January 2003 that a quarter of the genes in the yeast *Saccharomyces cerevisiae* are duplicated. Duplicate copies help buffer core functions against the effects of mutations. Interestingly, daughter genes within a duplicate pair can functionally compensate for parental genes, even when they are highly divergent. The effect of disabling 'singleton' genes that lack duplicated daughters is therefore more profound than disabling those that have given rise to gene families. Duplication can be more extensive. Eric Lander, at Harvard University, showed in April 2004 that the yeast *Saccharomyces cerevisiae* arose as the result of whole-genome duplication of the ancestral yeast *Kluyveromyces waltii*. This suggests that at least in some organisms, whole-genome duplication followed by massive gene loss and specialization has been an important mechanism of genomic reprogramming.

Examination of collections of junk sequences have identified non-coding sequences, known as conserved non-genic sequences (CNGs), that are remarkably conserved between different mammalian species. In some cases CNGs are more fully conserved than genes. The nucleotide substitution rate of the most highly conserved CNGs is less than half that of protein-coding regions. One explanation for this is that CNGs represent non-coding RNA (ncRNA) genes, genes that exclusively encode regulatory miRNAs – rather than the mRNAs, which are translated into proteins. It has been estimated that the human genome contains around 60,000 CNGs, more than twice the number of protein-coding genes. These account for nearly 1% of the human genome. Their function, however, remains a mystery. Comparative genomic studies have uncovered still further mechanisms of genomic reprogramming. Studies of three different species of yeast by Manolis Kellis, in May 2003, uncovered a mechanism that the authors called 'genomic churning', which is restricted to the ends of chromosomes, known as telomeres. Their results suggest that telomeric regions are a crucible in which genomic change is forged. They move back and forth between chromosomes carrying pieces of genes with them. Some of these combine to create new genes.

But in some cases genetic hardware conveys properties that are impossible to achieve by modifying software alone. Small modifications to the sequences of genes can have profound effects on the proteins they encode and may help define the identity of a species. The modifications needed to alter morphology dramatically can be far more trivial than one might have imagined. An example of this is found in the evolution of modern maize, *Zea mays* (*mays*), from its distant relative, the inedible weed teosinte, *Zea mays* (*parviglumis*).

The domestication of maize from wild grass was arguably mankind's first and most successful attempt at genetic engineering, and was achieved using intensive breeding. Modern maize is thought to have originated 6,000 years ago in Mexico, following a breeding programme starting around 9,000 years ago. The domestication of corn had profound social consequences, making food production much more efficient. This paved the way for complex communities in which only a small proportion of the population farmed, freeing the rest to do other things. The genetically modified maize produced by prehistoric farmers rapidly became a global staple. Unlike the frugal ears of teosinte, those of modern corn are huge and generously packed with kernels filled with protein, starch and oil. Intensive breeding has resulted in corn losing the ability to disperse its seeds and it is now dependent on humans for its survival. So what were the changes that underwrote this morphological transition, turning a useless weed into the human dietary staple? How was the genome of an ancient wild grass reprogrammed to make corn?

Studies in the 1990s traced the differences between maize and teosinte to only five genomic regions. In two of these the differences were attributable to alternative versions of a single gene. In one case the gene responsible was *tga1* (teosinte glume architecture 1), in the other *tb1* (teosinte branched 1). The *tga1* gene controls the nature of the armour-like fruitcase or 'glume' that surrounds each kernel. In teosinte the glume is stone-like, enabling kernels to survive passage through an animal's digestive system; in maize it is thin and barely develops. This modification makes the kernels soft, exposed and edible. The *tb1* gene, on the other hand, is largely responsible for the morphological differences between the plants. The version of *tb1* in teosinte produces multiple side branches, whereas the mutant version in corn suppresses the growth of lateral shoots. The result is an unbranched plant with a single cob. Another mutation, of a gene called *pbf* (prolamin box binding factor), increases the amount of seed storage protein in each kernel. Yet another mutation, of a gene called *su1* (sugary 1) changes the texture of the kernels. Modern corn is thus an assembly of innovations, multiple genetic modifications having accumulated in a single plant. Such constellations of small changes can in this way profoundly alter an organism's nature and may result in significant evolutionary leaps.

A further example of the power conveyed by individual bits of protein-coding hardware is found in the life cycle of leathery little creatures known as tunicates. One of these, a species called *Mogula occulta*, is found in the ocean off the city of Roscoff, in France. As adults they live like mussels, attached to shells or rocks, where they filter food through long siphons. Larval tunicates are very different, looking more like tadpoles than the quiescent adult rock suckers. They have a long dorsal nerve, a flexible rod of support cells (known as a notochord), a tail, and skeletal muscles that propel them across the shallow tidal flats. Remarkably, as they develop into adults they lose their notochord and tail. The presence of a backbone or notochord is a defining feature of vertebrate anatomy, so its loss effectively turns back the evolutionary clock, reverting the body plan of maturing tunicates to a prototypical invertebrate design.

Working at the University of California, William Jeffrey and Billie J. Swalla showed in November 1996 that the loss and gain of these vertebrate-like characteristics is controlled by a single piece of hardware, the *Manx* gene (named after a tailless cat). A closely related species, the tunicate *Mogula oculata*, lacks even an embryonic backbone, resulting from the disruption of *Manx*. Without *Manx* the whole set of characteristics is lost. These types of master control genes are likely to have played an important role in generating the unique body plans of each broad group (phylum) of living things. This includes chordates, the phylum containing animals with backbones. Until the discovery of the powers of the *Manx* gene, it was inconceivable that such profound anatomical modifications could be achieved so effortlessly. By switching the *Manx* gene off in embryos, the team re-created an evolutionary change that may historically have led to tailless tunicates. Indeed the acquisition of the *Manx* gene may have been of key importance in the origin of all chordates, including humans. The example of the *Manx* gene indicates how the other major transitions of evolution might have been achieved using master control genes.

The possibility of these kinds of dramatic leaps or 'saltatory transitions' between body plans is at odds with the classic Darwinian view of evolution as an incremental, gradualistic process resulting principally from the accumulation of point mutations. It is not yet evident which mechanism, gradual or saltatory, has been most important in the history of evolution, but it is likely that both have played their part. Macroevolution, the production of major new morphologies, has occurred by means of microevolution – that is, the acquisition of minor and occasionally major changes to genomic hardware and software. But in some cases the slightest of alterations result in sudden and dramatic morphological transitions, rather than incremental changes.

Some organisms have evolved mechanisms for buffering the effects of mutations to their genes. This allows genes to be altered without the intermediate and sometimes harmful effects of modifications becoming apparent. It also allows changes to accumulate in several different genes before revealing the morphological effects of the entire set. In so doing it facilitates rapid evolutionary change at times of environmental stress, generating a reservoir or stockpile of potentially useful genetic changes. This genetic 'capacitance' phenomenon was discovered by Suzanne Rutherford and Susan Lindquist in November 1998.

Under normal circumstances, the heat-shock protein *HSP90* helps protect and stabilize mature proteins. It also chaperones immature proteins during the 'folding' process when they adopt their three-dimensional shape. If a protein involved in fly development has been mutated in a way that altered its ability to fold correctly, *HSP90* might assist it. But when cells experience stress, from high temperatures, say, or lack of oxygen, *HSP90* proteins are overwhelmed by the sheer number of altered proteins, compromising the buffering capacity and preventing *HSP90* from carrying out its usual functions. In such emergency situations the mutant proteins, whose activities are usually masked, display their effects, both good and bad. If the outcomes are beneficial, the flies will survive and the mutations will either be fixed or remain as a hidden stockpile of future survival strategies. Mutated genetic hardware can as a result be stored so that their effects are hidden.

Rutherford and Lindquist began to suspect that *HSP90* might buffer mutations when they noticed that flies with

5. REPROGRAMMING LIFE

disabled *HSP90* genes had a variety of developmental abnormalities. These included misshapen wings and legs, abnormal eyes, faces, abdomens and bristles, and other oddities. When flies were fed with a substance that suppressed *HSP90* activity, 8% or more of the flies became deformed. The same phenomenon was observed when the temperature was raised a few degrees above or below the optimal value. *HSP90* can in this way shape an organism's evolutionary potential, radically altering its appearance or behaviour on the basis of the environmental context. The importance of this mechanism in macro-evolution, is that it allows evolving genes to bypass sequences that might encode unfavourable versions of proteins, thereby avoiding their adverse properties. By buffering alterations to the proteins that orchestrate development – and, most likely, others as well – *HSP90* helps ensure its robustness. The generality of the genetic capacitance mechanism became clear in May 2002 when Rutherford and Lindquist observed a similar phenomenon in the plant *Arabidopsis thaliana*. When *HSP90* was disabled a wide variety of exotic forms were observed. Leaves normally positioned at right angles adopted a whirling dervish formation, the colour of some plants darkened, and roots grew towards the sky.

The question of whether evolution occurs in the incremental manner of Darwinian theory, or in the saltatory, punctuated manner championed by Niles Eldredge and Stephen Jay Gould, is especially important when interpreting the fossils that document the apparently sudden emergence of the major types of animal body plans. This Big Bang of animal evolution, beginning around 540 million years ago, occurred in a six to ten million year geological instant known as the 'Cambrian Explosion', a time of profound genomic reprogramming. As a result, the simple animal forms that had populated the Earth were supplanted by a complex variety of clawed, armoured, feathered and scaled creatures that were the precursors of all future animals. The multicellular body plans that evolved during this evolutionary frenzy were the blueprints of those seen today. Just as all modern automobiles are variations of Henry Ford's original prototype, evolutionary changes since the Cambrian have been tinkerings with these ancient designs: not a single new phylum has been produced since then. Before then, life's evolutionary engine appears to have been stuck on idle for around three billion years.

The anatomical differences between animal design plans forms the basis for the Linnaean classification system that is still in use today. All creatures are divided into one of five kingdoms: *animalia*, *plantae* and *fungi*, which are multicellular; and *monera* and *protista*, which contain simple and complex unicellular animals respectively. Kingdoms are further subdivided into phyla, each representing the major body plans within each kingdom. The kingdom *animalia*, containing all known animals, is subdivided into around thirty-two phyla. Each is recognized by its unique design plan. Arthropods (insects, spiders, lobsters) use an extended set of repeated body segments;

chordates (encompassing all vertebrates, including humans) have backbones; molluscs (clams, squids, snails, starfish and sea urchins) have hard shells, soft, non-segmented bodies and typically possess a muscular foot. Each phylum is subdivided into groups. Arthropods, for example, are divided into uniramia (insects, centipedes and millipedes), chelicerata (spiders, mites and scorpions), crustacea (crabs, lobsters and shrimps) and trilobita (including the now extinct trilobites).

The apparently explosive nature of the evolutionary events in the early Cambrian present a significant problem for Darwin's gradualistic model of evolution. Ten million years is a very short time in evolutionary history. The sudden appearance of complex animal forms – including two chordates, *Pikaia* and *Yunnanozoon*, one of which may have been ancestral to humans – indicates that the tempo of evolution may have dramatically increased during this period. Darwin himself was puzzled, stating in *The Origin of Species* that 'the case at present must remain inexplicable; and may be truly urged as a valid argument against the views here entertained'. It is possible that the apparent abruptness with which complex life arose is simply an artefact arising from the fact that the soft bodies of the earlier intermediate creatures did not fossilize and left no trace. This was Darwin's preferred explanation. It is nevertheless possible that the apparently punctuated nature of evolution during the Cambrian did have a gradualistic basis. There is evidence that a phylogenetic 'fuse' of gradually accumulated genetic changes was lit long before, in the Precambrian era, when genomes tentatively experimented with complex form.

The reprogramming tricks available to genomes would certainly enable them to accomplish sudden, saltatory change. But whether evolutionary processes really are punctuated is another question. The fossil records of more recent evolutionary events provide a way of addressing this question. The palaeontologist Alan Cheetham studied the fossilized remains of coral-like animals known as bryozoa, hoping to ascertain the pace at which new species arose over the last fifteen million years. He was surprised to find that individual species appeared to remain unchanged for millions of years, but – in a geological instant of around 100,000 years – suddenly gave rise to new species. This un-Darwinian abruptness following long periods of stability appears, at least in this instance, to support the notion of punctuated evolution. Timothy Collins of the University of Michigan also found evidence for punctuated change in the evolution of coastal Californian snails over the last twenty million years. But other evidence – such as Peter Sheldon's study of Welsh trilobites – supports the gradualist paradigm. A study of the real-time evolution of bacterial cell size by Richard Lenski, however, suggests a punctuated model. Such conflicting results suggest that the processes by which genomes are reprogrammed may involve either kind of evolution. Indeed, the evolution of life on Earth may have utilized a judicious mix of both punctuated and gradualistic processes.

6

Making creatures from scratch

The 9 July 1846 edition of the Illustrated London News carried a description of a private view of a contraption at the Egyptian Hall in Piccadilly: 'one of the most extraordinary pieces of mechanism ever exhibited'. The device in question was the Wonderful Talking Machine, conceived and constructed by the sixty-year-old Professor Joseph Faber from Vienna, who had worked anonymously for many years as a land surveyor to the Emperor of Austria. Timid, awkward and depressive, Faber had spent the best part of his life building his machine, also known as the Euphonia; his desperation at his poor reception in Vienna and America had caused him to destroy an earlier version. His automaton, which resembled a life-sized Turk, lay in a reclining position supported by pillows. Fashioned from caoutchouc, or india-rubber, the Euphonia mimicked human speech organs and featured a jaw, tongue and throat, whose shape could be altered to produce a variety of different sounds. It was controlled not by tendons and muscles but by sixteen levers, and was vitalized by artificial lungs that consisted of a set of bellows operated by a foot pedal. The levers converged on a keyboard that produced elementary sounds: combinations of key presses could apparently generate 'every word in all European languages'. The *Illustrated London News* reported that the machine had distinctly pronounced the words 'Ehrenbreitstein', 'Jungfräulich', 'Philadelphia', 'thwart', and 'God bless Queen Victoria' – after which 'it concluded with a hurrah, and then laughed loudly'. Its virtuosity was not restricted to speech. The Euphonia also treated the astonished listeners to a perfect rendition of 'God Save the Queen'.

An article in the *Times*, on 12 August, was equally enthusiastic, insisting that it was 'almost a duty of all who can afford to see it, to gratify at once their curiosity, and show their encouragement to genius'. *Punch* magazine, alluding to the machine's rendition of the national anthem, described the automaton as 'indisputably the most popular singer of the day'. It also suggested that if 'combined with Mr Babbage's Calculating Machine' this 'ingenious union of wind and Indian-rubber ... might replace, with perfect propriety, a Chancellor of the Exchequer, or a mathematical Lecturer at the Universities'. Although critical of the machine's laugh, which was 'hard, dry and artificial' rather than 'hearty and jovial', *Punch* was most impressed by its hiss, which it felt was 'perfect', concluding that 'perhaps the fact suggests to the bene- volent mind the moral that hissing is the very easiest occupation of life'. But the talking machine, despite an endorsement from the Duke of Wellington, did not bring Faber the lasting fame and recognition he longed for: he

6. MAKING CREATURES FROM SCRATCH

was soon forgotten. The machine was displayed in a small theatre in Paris and then consigned to oblivion.

Faber's attempt at creating artificial life was one of the most technologically advanced the world had ever seen. But, like all its predecessors, his machine was not autonomous: it lacked the defining characteristics of living things. In August 1846 the *Illustrated London News* tempered its enthusiasm with a dose of sobriety: 'this is, like all similar attempts that have preceded it, only an approximation, though a nearer approximation, to the thing proposed'. Faber's contraption had been anticipated by the thirteenth-century philosopher Friar Roger Bacon, whose achievements are described in J. Freind's *The History of Physick* (1750). 'He, as we are told, could make a flying Chariot, and had an art of putting Statues into motion, and producing articulate sounds out of a Brazen head: and this not by any Magical power, but by one much superior, that of Philosophy and Nature, which can do such things, to use his own expressions, as the ignorant think Miracles.'

There had been other attempts at imitating life with machinery. In 400 BC Archytas of Tarentum built a wooden pigeon that simulated flight with the assistance of a water jet. To greet the Emperor Maximilian on his entry to Nuremberg in 1470, Johannes Muller was said to have made a flying artificial eagle that stretched its wings and bowed. But the most famous and accomplished automaton was the mechanical duck built in the early 1740s by Jacques de Vaucanson. Made from gold-plated copper,

Vaucanson's machine not only looked and moved like a duck, but imitated much of its internal anatomy. Each wing contained more than four hundred articulated parts. The machine also quacked and ate food from its inventor's hand. A 'chemical laboratory' had been installed inside its stomach to enable it to digest material introduced into its beak. Vaucanson admitted the machine's limitations: 'I do not pretend,' he said, 'to offer this digestion as a perfect digestion, capable of producing blood and nutritional elements for the animal's continuing health.'

Although the design of these automata incorporated rudimentary aspects of the logic of living things, none was even close to being alive. Life is notoriously difficult to define, but it is generally agreed that such characteristics as the ability for self-regeneration, self-replication and evolution are essential. The key property of living things, however, is that they carry an internal description of themselves written in the DNA code of their genes. This microcode, passed on from generation to generation, is the database from which the structures and behaviours of all living things are computed. No other physical system has anything of the kind. There is no internal description of weather, for instance, that is physically separable from the weather itself. The existence of DNA gives a system something radically new: the means to store, modify and manipulate information. It also dictates the construction of organs such as brains, which are able to generate abstract representations of the physical world and, in more

6. MAKING CREATURES FROM SCRATCH

advanced organisms, of the social, mental and cultural environment as well. The desire to understand life and create new life is as vigorous now as it has ever been; the difference is that such grand designs are close to being within our grasp.

Biochemical and genetic studies show that the computational model of living things is a good starting-point for such endeavours. From this perspective, the logical form of life is abstracted into genetic circuit diagrams that interconnect to generate living things. The computational model may not ultimately prove to be correct. Gene networks have emergent properties that cannot be inferred from the behaviour of individual components. But the model is testable and can be investigated using artificial genetic circuits. The synthetic approach to life is the logical conclusion of the protracted analytic phase of biological discovery. Analytic biology attempts to understand life by breaking living machines down into their components; synthetic – or 'constructional' – biology gains insights by building living things from first principles. It is not necessarily important that the computational model is correct, so long as it is useful. Newton's laws have enabled the exploration of the Moon and Mars by allowing physical behaviour to be anticipated anywhere in the universe. But we should remember that – in spite of the success of Newtonian models – our understanding of the physical basis of gravity is still rudimentary. To model a phenomenon, and to produce accurate predictions on the basis of that model, is not the same thing as understanding it fully.

The foundations of synthetic biology were established on 20 October 1995, when Craig Venter, working at the Institute for Genomic Research (TIGR), published the complete genome sequence of *Mycoplasma genitalium*. These tiny bacteria live parasitically in the genital and respiratory tracts of humans. *M. genitalium* was the second ever bacterial genome to be sequenced, but was especially important since it has one of the smallest genomes of any known cellular organism: its parasitic lifestyle has led to its genome being stripped down to bare essentials. It has just 517 genes, 480 of which encode proteins; the other 37 encode RNA molecules. Its simplicity offers the possibility of defining the minimal gene set compatible with life. By December 1999, Venter and his colleague Clyde Hutchinson had deleted many of the protein-encoding genes in *M. genitalium*, showing that only around 300 were essential for survival. But a minimal gene set is not the same as a minimal genome, since complete genomes also include non-coding elements. An entirely new experiment of the same kind would be needed to determine the minimal non-coding DNA complement essential for the correct functioning of *M. genitalium*. In order to define a minimal gene set, Arcady Mushegian and Eugene Koonin compared the gene complement of *M. genitalium* with that of a different species of bacteria, *Haemophilus influenzae*, and identified around 240 shared genes. They used this to estimate that

6. MAKING CREATURES FROM SCRATCH

the minimal gene set necessary for life is likely to contain around 250 genes. Any genome – a clam, tree, leopard or human – can be thought of as an expansion of the minimal set. Mushegian and Koonin independently estimated that the 'ancestral set', from which all modern gene sets arose, contained around 128 genes. This minimal gene set might provide clues as to how life first originated.

All such 'top down' methods, whether experimental or theoretical, are hampered by a problem: they cannot be used to test the prediction that life can be booted up with a minimal gene set. Nor can they tell us precisely what combinations of genes would be required for basic life. To test these intriguing hypotheses, a completely new, 'bottom up', approach to genetics is required. One such method involves synthesizing artificial genomes that contain candidate gene sets and 'running' them in the real world to see which ones work. Non-coding DNA can then be added to identify regulatory sequences essential for the genome to function. The skeleton of a minimal genome provides a foundation on which larger genomes can be built.

As a first step to creating artificial genomes of cellular organisms, Craig Venter teamed up with Hamilton Smith to synthesize an artificial viral genome. They cobbled together a bacteriophage – around one hundred times smaller than the 580,000 nucleotide long genome of the *M. genitalium* bacteria – in only two weeks. Their result, published in November 2003, was achieved by synthesizing small pieces of DNA and piecing them together by matching overlapping ends. Venter and Hamilton Smith were not the first people to stitch together an artificial genome. In August 2002 Eckard Wimmer at the State University of New York at Stonybrook announced that he had synthesized an artificial poliovirus. But this had taken him three years.

Of course, viruses are much simpler creatures than bacteria and all other cellular life-forms, which means that their genomes are more straightforward to construct. The leap from building an artificial viral genome to building even bacterial genomes is considerable, and presents many technological challenges. In particular, there are problems associated with structures such as cell membranes and organelles such as nuclei and mitochondria that house genomes. Artificial genomes may also need to be 'booted up' (though it is possible that this might occur spontaneously); otherwise, synthetic genomes would lie inactive and impotent within cells. Providing the appropriate genetic software in terms of regulatory sequences is also essential. Perhaps the most difficult issue, however, is determining which genes to include in an artificial genome. It would be tiresome to have to synthesize every possible combination of genes, to find out which cells survived. Fortunately, bacterial genomes are organized into functional 'cassettes' of genes that perform similar functions and are clustered together. The *lac* cassette, or 'operon', for example, contains three functionally related genes – *lacZ*, *lacY* and *lacA* – required for the metabolism of lactose and related sugars. This means that, at least in

6. MAKING CREATURES FROM SCRATCH

the case of bacteria, there is a practical alternative to assembling combinations of genes from scratch: different gene cassettes can simply be spliced together.

Viral genomes exist autonomously within the cells that they infect, but in order to test-drive a newly synthesized bacterial genome it would have to be inserted or 'downloaded' into a pre-existing bacterium whose own genome had been excised. This microbial chassis would provide the basic structure and microenvironment within which the genome would function. In the future, however, it should be possible to construct artificial cells into which synthetic genomes can be inserted. The synthesis of eukaryotic (higher organism) genomes is a much tougher proposition than the synthesis of prokaryotic (bacterial) genomes. This is not only because eukaryotic genomes are far larger and are not organized into cassettes, but because they include significant amounts of non-coding DNA, some of which is essential for gene function and genome stability. The effects of inserting non-coding DNA into a synthetic genome should be computable, but it will be many years before these principles are understood. Until that time, the background non-coding sequences suitable for inclusion in synthetic genomes will be determined largely by trial and error. Synthetic genetic circuits will be designed, constructed and tested. If they work, the configuration will reveal the underlying logical principles; if they do not, they can be reconfigured until they do.

But when constructing synthetic organisms many other engineering issues need to be considered. Genetic systems need to be robust, to prevent the system 'crashing' or losing vital functions through small changes in their operating parameters or mutations in gene or regulatory sequences. If synthetic organisms are to mimic natural organisms fully, they also need to be 'evolvable': they must have the capacity to mutate in order to respond to new environmental contingencies. One day synthetic cells will be part of the bio-designer's standard tool kit. Tiny programmable biological computers and the biochemical modular subroutines from which they are composed will be available as 'off the shelf' components, like the transistors and logic boards of micro-electronic devices.

Synthetic genomes will help address academic questions. But there are practical advantages too: artificially constructed organisms could be programmed – by adjusting their hardware and software – for a variety of uses. Living devices will join mankind's other artefacts. Synthetic bacteria could, for instance, be programmed to destroy environmental pollutants. They could also substitute for fossil fuels and be used to generate energy; or to metabolize the excess atmospheric carbon dioxide responsible for global warming. The downloading of synthetic genomes into cells turns them into 'nano-robots'. The complete reprogramming of genomes through artificial design and synthesis is not the only way of producing new cellular behaviours, however. In April 2004 Frances Arnold and her colleagues at the California Institute of Technology showed that the behaviour of *Escherichia coli* populations could be reprogrammed by the introduction of two tiny genetic circuits that functioned alongside the parental genomes. Mathematical models indicated that the artificial circuits would alter cell population dynamics, encouraging the bacterial population to maintain lower cell densities than normal, and when artificial gene circuits were downloaded into the bacteria, this is precisely what happened. The density of cells could be altered by tinkering with the properties of the artificial gene circuits.

Results of this kind support the notion that genomes are inherently 'computable', indicating that if all the necessary information were to be assembled, the network dynamics – or 'interactome' – and behaviour of individual cells, populations of cells and whole organisms could be modelled in detail. Once the computational logic of cells has been understood, new genomes could be synthesized on the basis of logical principles: 'dry runs' could be conducted in computers before 'wet runs' are attempted using the DNA and proteins of living things. The first step towards modelling the behaviour of individual cells has already been made by Masaru Tomita at Keio University in Japan. His E-cell software constructs a virtual cell composed of 127 genes based on those of *Mycoplasma genitalium*. His cell 'lives' on a computer screen, taking up virtual glucose from a virtual culture medium, synthesizing virtual proteins and generating virtual waste. Experimental manipulations, such as starving the cell of its virtual glucose, have generated surprising results –

findings that could not have been predicted from studies of isolated gene circuits.

Efforts to model the logic and behaviour of 'intelligent' cellular systems have now been formalized, in the shape of the *E. Coli* Model Cell Consortium – an undertaking that dwarfs the Human Genome Project in both complexity and importance. The genome project merely provided a complete catalogue of the parts that make up a cell; the Model Cell Consortium plans to produce a definitive model of how, in the bacterium *E. coli*, these parts fit together. Once the inter-relationships between the components of simple cells are understood, it will be possible to model more complex cells, and eventually those of humans. Attempts to model development will follow. It should eventually be possible to infer the morphology and general nature of synthetic organisms from their DNA sequences and epigenomes alone.

The immense complexity of the logical rules that underlie the control components of individual 'smart' genes is evident in the *Endo16* gene, found in the sea urchin *Strongylocentrotus purpuratus*. *Endo16* is expressed in cells that become the gut of the developing sea urchin embryo. It has thirty different control regions in its promoter; these are responsive to at least thirteen different regulatory factors. The control regions act as 'on-off' switches and 'volume controls', operating on the basis of simple logical principles. The complexity of the control system enables *Endo16* gene expression to be sensitive to a wide range of combinations of environmental signals; it integrates them like a miniature analogue computer circuit, producing a range of appropriate output responses. The logical rules it operates under are of the simple Boolean type. In a NOT IF gate, for example, the gene will only be switched on in the absence of a specific protein input. In an AND gate, the gene will be switched on only if all inputs are present. The 'program' that runs on this tiny computer is hard-wired into the DNA regulatory elements; the inputs are the molecules that bind to the regulatory elements; the output is the resulting pattern and level of gene expression. A module consists of several such genes interacting to produce a specific function, with the expression of each component gene being controlled by the operation of Boolean logic.

The RNA and protein molecules encoded by a given genome can be clustered into groups of molecules that co-operate towards a specific functional end. Living things are composed of a series of modular units that together give each organism its particular logical form. Each genetic module, which can be thought of as a basic building block of cellular organization, consists of a related group of genes and the proteins they encode: it functions as a biochemical machine with liquid parts. Each performs delineated subroutines of the genetic computer. Stuart Kim and his colleagues at the Stanford University School of Medicine have provided evidence for the module hypothesis by demonstrating in October 2003 that the patterns in which groups of genes are switched on and off is conserved in a wide range of different species, from humans and fruit flies, to worms and yeast.

The agency of living things emerges from the integration of large numbers of these subroutines which seamlessly interlink to produce their dynamic properties. The extent to which modules are interconnected varies: some are highly integrated into the network; others are relatively autonomous. In different creatures, variations in the connectivity between interacting modules lead to different network dynamics. The difference between a hedgehog and a cod, a zebra and a lion, a parrot and a penguin, or a stickleback and a scorpion, is simply that they have different hardware and software. Their shared biochemistry and logical principles demonstrate the intrinsic continuity and unity between all Earth's species.

The invention and modification of new genetic modules by natural evolutionary processes has formed the logical basis for life on Earth. Humans are organically connected with all other species: we differ from other creatures – including our closest relatives, the African apes – only in our unique inventory of hardware components and the software programs that regulate them. All our characteristics, including consciousness, are generated by the agency of genetic minicomputers inside our cells. But there is no reason why synthetic biology should restrict itself to naturally occurring modules. It is likely that natural evolution has discovered only a fraction of the possible types and combinations of modules. Through the use of synthetic genomes, the circuitry of existing species

6. MAKING CREATURES FROM SCRATCH

can be mixed and matched to produce completely new biological structures and behaviours. New modules conveying novel properties can also be engineered, 'from the circuit board up'.

The construction of genomes from scratch is an endeavour in a different league from earlier modifications, which involved the introduction of only small numbers of genes into pre-existing genomes. This does not mean that simple genetic modifications are unimportant. The presence of a new piece of genetic hardware can have important consequences. 'Golden rice', for example, is a genetically modified strain that contains the beta-carotene gene and may help combat vitamin A deficiency in developing countries. But the addition of a single gene can have more fundamental effects. K.E. Apt showed in June 2001 that the metabolism of the microalga *Phaeodactylum tricornutum* could be completely reprogrammed by the artificial insertion of a gene called *glut1*. This tiny creature usually relies on sunlight for energy, converting it by photosynthesis; but the introduction of *glut1*, which is a glucose transporter, enables it to thrive on exogenous glucose in the absence of light.

Genome sequencing projects have been an important step on the way to synthesizing genomes: gene inventories can be annotated, by allocating each gene and its products to a functional class. The genes and proteins in the yeast *S. cerevisiae*, for example, are divided into 17 categories. These include transcription factors that control gene expression (152 genes), enzymes involved in the metabolism of amino acids (72 genes), and enzymes involved in glucose metabolism (16 genes). Within each broad category, individual metabolic pathways constitute the subroutines of the major functional type. Amino acid metabolism, for instance, can be broken down into 20 different amino acid synthesizing modules, one for each amino acid. These modules would form the basis of any artificial genome coding for proteins composed of the 20 standard amino acids. Such essential modules are expected to be common to virtually all living things, although their precise wiring and configuration varies from species to species, modified by evolution in each case for a different functional emphasis.

Genes may be divided into 'families' and 'super-families' on the basis of distinguishing motifs in their DNA sequence, many of which are essential in delivering the function of a module. The extent to which each species utilizes particular gene families varies greatly. Only 1% of the genes of the fruit fly belong to the immunoglobulin supergene family (used, among many other things, to make antibodies); in mice and humans this family has expanded considerably, accounting for 2.7% and 2.8% respectively of the total gene complement. In nematode worms, on the other hand, the proportion is only 0.4%. The pufferfish *Fugu* lies somewhere in between, at 1.7%. But within any given gene family there is much diversity. The immunoglobulin supergene family contains considerable functional complexity: the same protein structure is adapted to generate an extremely varied molecular toolkit.

6. MAKING CREATURES FROM SCRATCH

The same genes can also differ greatly between species, and between individuals within a species. With a very few modifications, the haemoglobin gene can enable llamas to live at high altitudes and crocodiles to stay under water for extended durations. The haemoglobin gene also differs between humans: sickle cell mutant forms confer resistance to malaria; other mutants are purely detrimental.

The new synthetic biology will combine existing and artificial modules to generate new functions and cellular behaviours. Eventually, it will be able to predict the behaviours that any particular combination of modules produces, and therefore the type of creature that will result. The predictive science of synthetic genomics – and the generation of new life forms from first principles – will be facilitated once the genome sequences of all existing living creatures have been determined. But this synthetic genomic science will require extensive knowledge that goes far beyond the information stored in genes. It will have to take into account non-coding elements (such as promoters, enhancers and transposons), epigenetic information, and extra-genetic information acquired by learning and cultural transmission. It will also, in some species, include information about proteins present in eggs that are essential for kick-starting development.

It is now possible to observe genetic programs in action, thanks to what is known as gene array technology. The technique involves immobilizing a small portion of each gene in the genome under study, on a two dimensional surface. Fluorescent-dyed mRNA molecules are then introduced, and bind to the sequence they correspond to where a gene is switched on. A scanner sensitive to fluorescence detects the binding events. The amount of fluorescence indicates the extent to which any particular gene is switched on or off; a computer program translates the different levels of gene expression into easily observed colours. In October 2003 G.E. Robinson and his colleagues at the University of Illinois used this method to examine gene expression programs in the honeybee *Apis mellifera*. In the last five to seven weeks of their life, honey bees undergo behavioural modifications: they change from being home-loving hive workers, helping to nurse young bees, into extravert foragers, flying frenetically in search of pollen and nectar. They can travel up to several kilometres in a variety of weather conditions. After studying gene expression patterns in the brains of around sixty individual bees, Robinson identified a set of fifty genes whose patterns of expression could be used to predict whether each bee was a hive-dweller or a forager.

Using a similar approach, Kevin White studied the expression of 4,028 fruit fly genes during different stages of development and identified gene expression programs corresponding to different developmental stages. In a study published in October 1997, Patrick Brown at Stanford University used an array containing yeast genes to examine the genetic basis of the metabolic reprogramming that occurs when yeasts shift from anaerobic

6. MAKING CREATURES FROM SCRATCH

metabolism (fermentation) to aerobic metabolism (respiration). He was able to define the gene expression programs associated with each mode of metabolic behaviour. Comprehensive maps of gene expression programs, corresponding to different stages of development, activation and to different behaviours, will allow the behaviour of synthetic organisms to be precisely dictated. Artificial bees, for example, could be prevented from running foraging programs. This would prevent their natural age-related change in behaviour from occurring: they would remain hive-bound for their entire lives. Subject to certain constraints, synthetic organisms could be endowed with any number of properties.

Although it is possible that the artificial synthesis of genomes will proceed successfully using basic engineering principles alone, applied on a gene-by-gene or module-by-module basis, complete metabolisms – consisting of entire networks of interconnected proteins and miRNA molecules – may have emergent properties that transcend the behaviour of the individual modules from which they are composed. Computer simulations of networks of simple repetitive units that individually behave according to Boolean logical rules are known to generate emergent 'non-linear' behaviours. These 'global' behaviours are dictated by the number of interconnections between each node and by the Boolean logic employed at each node.

It is likely that the topology of metabolic networks will affect their dynamical behaviour, as well as other properties

– such as their robustness and their capacity to evolve. Stanislas Leibler has provided evidence for the importance of connectivity in biological networks by synthesizing miniature genetic networks consisting of combinations of three genes and five promoters. Using these starting components, 125 networks are possible. The behaviour of the different logical circuits produced was found to be a function both of their connectivity and the components utilized. The remarkable result, however, was the diversity of complex behaviours that could be produced, using only a handful of genetic elements. It is likely that evolution has tuned metabolic networks by adjusting the connectivity and rules operating at each node in order to foster some emergent properties whilst suppressing others. Such phenomena will have significant implications for the design of synthetic organisms, and may limit what is attainable. In February 2004 A.L. Barabasi, at the University of Notre Dame in Indiana, performed computer simulations of the bacteria *E. coli* and showed that the flux of metabolic activity in most relevant modules was low: the overall activity of the metabolism was dominated by the activity of only a small number of reactions. The varying flux was a direct consequence of the network topology, which channelled numerous small fluxes into high-flux pathways.

Although it is still easier to use pre-existing or 'natural' hardware and software components for the design and construction of synthetic life, there is no reason why this should be the case in the future. Artificial genes could be

6. MAKING CREATURES FROM SCRATCH

built, including those not used by modern organisms. It may be possible to rescue the genes of extinct organisms and to combine them with those of contemporary organisms. The construction of artificial proteins would give organisms new properties not previously seen in nature. The palette of the synthetic biologist will increase once the genomes of all known organisms have been sequenced. These could then be scoured for interesting genes and modules: synthetic organisms could be constructed that are quite unlike anything that has ever lived before.

7

The limits of possibility

According to the author of an article that appeared in the *Times* newspaper on 12 February 1866, 'nothing to the mind of a young lady could possibly be more delightful than the life of a stage fairy'. Fairy garments were 'seemingly fashioned from imponderable gold or silver'; they spent their time 'screwed down or suspended in picturesque attitudes' or 'floating through the air surrounded by gorgeous fabrics'. The real day-to-day life of these insect-like and 'smartly-dressed young ladies ... supposed to represent beneficent beings endowed with supernatural capabilities' was somewhat different. The London theatrical Christmas lasted only two months; once the pantomimes were over and March had come, the mortal fairies faded away. Although there were a number of provincial engagements to sustain them in the summer, autumn contained 'a series of ugly weeks', during which the fairies gained 'not one farthing' for their fairyhood.

Victorians were obsessed with fairies. They were essential to the work of many painters, poets, novelists – and even scientists. The belief in their existence was widespread. Fairy sightings were commonplace throughout the eighteenth and nineteenth centuries. The chemist Robert

7. THE LIMITS OF POSSIBILITY

Boyle gave his blessing to a project initiated by the Reverend Robert Kirk that aimed to provide empirical confirmation of Christian spirituality by assembling reported sightings of fairies in the highlands of Scotland. There are documented accounts of sightings in the seventeenth century too. In a letter dated 1st May 1696 Moses Pitt recounts the story of Ann Jeffries, who, one day in 1645, was 'knitting in the Arbour in our garden' when 'there came over the garden-hedge to her ... six persons of a small stature, all clothed in green, which she called fairies: upon which she was so frightened, that she fell into a kind of convulsion-fit'. These diaphanous winged creatures embodied the anxieties of Victorian romantics: anxiety at the destruction of the familiar and the viciousness with which custom and tradition were being corrupted by the soulless dust of modernity. As science stripped the world of its mystery, enchanted landscapes peopled with wondrous beings promised an existence beyond the purely rational, scientific and secular. They offered a haven suffused with the enchantment that was daily disappearing from people's lives, and as such were part of the anti-rationalist movement dubbed by Isaiah Berlin as 'counter-Enlightenment'.

Although sensitive to slights, for which they invoked severe retributions, fairies were portrayed as being indifferent to human affairs, going about their business peacefully until disturbed. They always had a human or human-like morphology, but were also said to have the power to change their appearance and to become invisible.

Some accounts described fairies as being as tall as three feet, but they were generally no larger than the palm of a human hand. The insect-wings synonymous with modern depictions only became a characteristic feature following the artistic portrayals of the nineteenth century – seen, for instance, in the oil paintings of John Anster Fitzgerald. But fairies were not appreciated by everyone. At the 1821 Royal Academy summer exhibition the critic T.G. Wainewright claimed to have overheard the following exchange between two ladies viewing one of Henry Fuseli's fairy paintings. 'Whose is that?' 'Fuseli's.' 'What a frightful thing! I hate his fancies of fairies and spirits and nonsense. One can't understand them … It's foolish to paint things which nobody ever saw, for how is one to know if they're ever right?' Wainewright may, of course, have made up the dialogue, using the comments as a device to attack a frivolous public who had failed to appreciate Fuseli's work. He did at least appreciate that 'things which nobody ever saw' had their own reality.

Men tended to appreciate fairies more than women. Fairies, tiny and beautiful, came to symbolize the idealized Victorian woman. Magical, unavailable, delicate in constitution, diminutive, playful rather than earnest, they were the personification of feminine form and temperament. But they also represented every aspect of femininity that the newly emerging modern women were trying to escape from. In reaction to an unsympathetic male invention, Mary Braddon, in her novel *Lady Audley's Secret* (1863), uses the term 'fairy-like' to evoke an imperfect

7. THE LIMITS OF POSSIBILITY

character blighted by duplicity and falsehood. John Black, a more sober commentator, simply claimed that fairies filled the gaps in common people's empirical knowledge of the natural world. For some people, he wrote, 'the imagination blends itself with the reality, the wonderful with the natural' and 'the false with the true'.

The existence of fairies, real or imagined, raises some important questions. Might it be possible to build a fairy from first principles? What about centaurs, mermaids, dragons and unicorns? Are there any limits to what nature can achieve? Can these limitations be transcended by artificial means? Could we build humans with bat wings and octopus eyes? Is the organization of the natural world and the pattern of speciation and extinction that the Earth has witnessed inevitable in every situation? Is evolution programmed to deliver the same end points repeatedly? What if we were to rewind the tape of life and let it play once again: would humans necessarily be an outcome? Might the genetic systems of living things be constructed from chemicals other than DNA and RNA? Are proteins necessary for life? All these issues fall under the same category, and may be summarized as follows. Are there constraints on life which prevent the realization of certain clusters of characteristics in a single creature and that guide and delimit the allowed patterns of evolution, channelling them into certain preferred directions? The answer, of course, is that there are; and that life is not completely open-ended. But how far-reaching are these constraints?

Constraints may be divided into several broad types.

Physical and chemical constraints relate to fundamental engineering impossibilities. Basilisk lizards can run on water, but to achieve the same feat, humans would have to weigh around 4.6g. The design of the burrows of prairie dogs is constrained by the fact that they are ventilated according to the Bernoulli principle (as the speed of a gas or liquid increases, its pressure decreases). The tallest living organism on Earth is a giant redwood tree, *Sequoia sempervirens*, in northern California: at 112.7 metres high, it is roughly equivalent to a 30-storey building. But what limits the height of trees? Why are redwoods not as tall as a 60-storey building? It seems that the fundamental physical constraint limiting tree height is not mechanical stress but the problem of providing a water supply. Water rises to treetops by a process called transpiration: it is pulled through specialized water-conducting tissue known as xylem. This force overcomes gravity and friction, but at a certain threshold the water column can no longer endure the tension. At this point air bubbles are released that can severely damage the plant. When tensions are measured in redwoods they are found to be close to the maximal point, suggesting that trees are limited to around 130 metres in height. This does not necessarily mean that it would be impossible to design artificial water-conducting systems to replace the xylem, which might allow higher water columns to be tolerated.

An unusual constraint of this sort – whose significance remains something of a mystery – is the appearance of the

7. THE LIMITS OF POSSIBILITY

Fibonacci series in a wide variety of plants. Fibonacci was the nickname of the Italian mathematician Leonardo Pisano (1170-1250), who, among other things, was responsible for introducing the Arabic number system to Europe. In 1202 he solved a problem relating to the breeding of rabbits that required a number series (1, 1, 2, 3, 5, 8, 13, 21, 34, 55 ...) in which every member is the sum of the previous two. The series that now bears his name is found in the arrangement of a great many plant organs, though nobody knows why. The seeds of the sunflower *Helianthus annuus* and the leaves of cacti and succulents such as *Mammillaria myrtax* and *Sempervivum hybrida* are arranged in spirals; the numbers of leaves or seeds are consecutive Fibonacci numbers. The recurring Fibonacci patterns suggest some underlying constraint that is not yet understood. Ultimately, all aspects of life are in some way constrained by physical or chemical laws. Much of the complexity of living things is not encoded in genes at all, but is intrinsic to natural laws: genes merely channel these processes to specific ends. Viruses contain large numbers of identical proteins that self-assemble to produce the virus's coat. There are no coded instructions in the virus to control this process: this is 'information for free'. The laws of self-assembly and self-organization, whose program-free action can be seen in the complex morphology of structures such as snowflakes, are assumed as a given by genes, and are part of the basic palette of natural physical and chemical phenomena that genes exploit.

'Developmental constraints' result from the way in which organisms are constructed during development and the way in which body parts are interlinked. In some cases these constraints will prevent combinations of characteristics from occurring together. It may, for example, turn out to be impossible to devise a developmental program that could build intelligent prawns with horsehair, wings and lizard tongues. The horns of the dung beetle *Onthophagus*, which they use to fend off competing suitors, are massively out of proportion to their body size. One would imagine that evolution would select for progressively larger and larger horns – as in an arms race where the odds are continually raised. This does not in fact happen: horn size seems to have an upper limit. There is also an inverse correlation between horn size and eye size. Beetles with excessively big horns have tiny eyes, and are virtually blind; those with short horns have large eyes and relatively good vision. This trade-off between nearby structures appears to be a developmental constraint resulting from the fact that neighbouring structures compete for finite resources during development. Either characteristic can increase in size and complexity, but only at the expense of the other. Big horned beetles with large eyes and good vision would only be attainable if the development of eyes and horns could in some way be uncoupled. It may be that no alternative developmental mechanism is possible, in which case certain combinations of features would be unattainable, even in synthetic organisms.

A similar phenomenon is observed in animals, all of

7. THE LIMITS OF POSSIBILITY

which have either good vision or a good sense of smell, but not both. In humans, for example, around 60% of the 1,000 or so genes that encode receptors involved with smell have been decommissioned, and remain as inactivated pseudogenes. Mice and dogs, on the other hand, which are colour-blind, have decommissioned only around 20% of these genes. It seems that there is a trade-off between good vision and a good sense of smell. This constraint most likely arises from the fact that vision and olfaction both take up large amounts of space in the brain: there simply is not enough space to accommodate expansion in both departments. The price of the evolution of sharp colour vision in humans may have been the relative decline of a sense of smell.

Relative constraints, on the other hand, relate to organisms that could in principle be built but that would be unlikely to survive if constructed on Earth. It is possible to imagine genetic programmes capable of building butterflies with wing-spans the size of tennis courts, but such creatures would not survive independently, for reasons of efficiency and physical implausibility. Oversized and clumsy, they would be unlikely to be able to fly and would lie wretched and immobile until they starved. Of course in a different physical environment, such as the weightlessness of space, they might well endure – if they could feed themselves.

Inevitable constraints relate to structures that emerge organically as a result of the introduction of other features, even though their existence is not necessarily expected, required or even useful. Stephen Jay Gould and Richard Lewontin have provided a good example: the architectural feature known as the spandrel. Spandrels – tapering, triangular spaces formed by the intersection of two rounded arches at right angles – can be seen in San Marco in Venice, and are the inevitable architectural by-products of mounting a dome on rounded arches. Spandrels are not an intended design feature of the dome, but emerge unavoidably as intrinsic architectural constraints associated with realizing the primary feature, the dome. At San Marco the spandrels were put to good use, being decorated with ornate iconography, but such additional emergent features are not necessarily adaptive. Some, indeed, may be detrimental: you have to take the good with the bad.

Functional constraints reflect the fact that optimal designs may not be ideal in the real world. The human respiratory tract is a good example. The geometry and dimensions of the bronchial tree, through which air from the outside world is conducted into the alveoli of the lungs, are critically important in determining the efficiency of breathing. If the size and shape of the bronchi are non-optimal, as with asthmatics, the resistance to airflow increases dramatically, and breathing becomes laboured. Small variations in the size and geometry of airways can dramatically influence airflow. One might expect evolution to have fine-tuned the respiratory system to produce an optimal bronchial tree, but it has been shown that optimal bronchial trees are sensitive to

7. THE LIMITS OF POSSIBILITY

changes in various parameters and potentially dangerous. Actual bronchial trees therefore have a sub-optimal morphology, all the bronchi being a little larger than optimality demands. This is a necessary constraint that enables the system to have another very important property: robustness. Another type of constraint is the need for biological systems to retain the capacity for evolution, or 'evolvability'. Brittle structures that are not able to evolve stand little chance of long-term survival: it is likely that this sort of 'evolvability constraint' limits the types of structures that can be perpetuated across time.

Material constraints relate to the fact that living things use DNA molecules to encode their information, and RNA and proteins to execute it. Both technologies are immensely versatile, but their use necessarily imposes constraints on living things. The physical properties and chemistry of DNA, RNA and proteins put a limit on the types of structures and reactions that living things are able to realize. It remains to be seen whether living things can be made from alternative materials, but it ought to be possible – just as buildings may be constructed from breeze blocks, bricks or steel, each with its own unique set of physical, chemical and aesthetic properties. When life originated on Earth around 3.6 billion years ago, evolution selected DNA, RNA and proteins as the life technologies. It is possible that if the tape of life were rewound, other chemistries or alternative versions of the same technologies, with different properties, might emerge instead. It may be that life can only be realized in these chemistries, but it is unlikely. Organisms constructed from these conventional materials are suited for life on Earth, but if we wanted to populate the surface of other planets, alternative technologies might be needed.

Synthetic organisms could encode their information using different or modified software technologies in place of DNA and hardware technologies other than proteins to realize it. The overall logic of their operation, however, is likely to be conserved. There are several informational polymers that could substitute for DNA, though if they were to be useful such artificial software molecules would – like DNA – need to be repaired, maintained and replicated. In October 2003 Haibo Liu at the Department of Chemistry in Stanford University synthesized a new form of DNA in which the C, T, G and A nucleotides were replaced with larger molecules. The resulting artificial DNA was both larger (by 25%) and more stable than its natural counterpart. Liu's work shows how artificial genetic systems might be constructed. In September 1997 Jeffrey Moore and his colleagues at the University of Illinois built a new material made from phenylacetylene, which, like proteins, coils into a helix and forms three-dimensional shapes or 'folds'. Such molecules could in principle be used in place of proteins, and would have the advantage of being more stable. There are twelve nucleotides other than A, C, T and G that could combine to make a more conventional artificial DNA molecule. But evolution has selected the A, C, T and G nucleotides for use in

7. THE LIMITS OF POSSIBILITY

its software for reasons that probably relate to the error-correcting aspects of information transfer and their intrinsic compatibility. Use of the standard set most likely represents the result of selection rather than a chance event.

Another material constraint relates to the amino acid composition of proteins and the effect of this on their properties. One of the most intriguing questions of evolution is why the proteins of all living things, with two known exceptions, are assembled using combinations of only 20 different amino acids. The exceptions, which expand the richness of the genetic code, are the microbe *Methanosarcina barkeri*, which has modified its genetic code to include an extra amino acid called pyrrolysine; and a number of animals and bacteria that utilize the non-standard amino acid selenocysteine. In both, this is achieved by engineering the genetic code to redefine the meaning of one of the stop codons marking the end of genes. Organisms that exist in extreme environments, such as hyperthermophilic bacteria that live in boiling springs, adapt in a more conventional manner: not by the incorporation of new amino acids, but by dramatically modifying the amino acid composition of their proteins using the conventional set.

Although many organisms have non-standard amino acids in their proteins, these are usually made by modifying the standard set once the protein has been made. In two dramatic experiments, Peter Schultz and Philippe Marliere re-engineered the basic machinery that cells use to make proteins, creating the first truly autonomous synthetic life. They engineered the genetic code of an *Escherichia coli* bacterium to introduce a 21st unnatural amino acid into its proteins. These bugs had a biochemistry unlike anything previously found in nature. The expansion of the natural amino acid set by artificial means allows new physical, chemical and biological properties to be introduced into proteins. For over three billion years life has made do with 20, and in a very small number of cases 21 different amino acids; but in the future such constraints will not be necessary.

Historical constraints are different. They relate to situations in which the starting point makes it difficult or impossible to proceed to a particular end point, even when it might in principle be attainable by another route. If, for example, a cataclysmic event had made all life on Earth extinct, apart from gorillas, would it be possible, even in principle, to evolve a lobster from a gorilla? Lobsters have fundamentally different body plans from gorillas: so it is hard to believe that a natural process of evolution could transform them into lobsters, irrespective of the time available and the intensity of the selection pressures. The feasibility but impossibility of the realization of a lobster in this situation is an example of a historical constraint and results from the dependence of natural evolutionary processes on the starting material. Natural evolutionary processes are unable to redesign organisms from first principles. As with an aeroplane, which can be modified but has to fly whatever the modifications, evolution builds

7. THE LIMITS OF POSSIBILITY

on the past, modifying pre-existing structures and retaining a functioning creature throughout. Old structures and control systems remain in place, with new ones integrated into and layered over them.

Another example is summarized in Dollo's law: once a species loses a particular characteristic, the character elimination tends to be irreversible. Little is known about the cause of this, but it is thought to result from the fact that many characteristics result from the interactions of multiple genes. Such characteristics are often eliminated by the inactivation of only a single gene in the network. If the remaining genes serve no other functions, they are free to accumulate mutations; the pathway degenerates as the genes of the former network evolve new functions or are inactivated to become pseudogenes.

Another type of historical constraint relates to the restricted set of proteins used in living things, a mere fraction of the set of all possible proteins. There is a dense network of kinship relations between many proteins. Each family is built around a particular three-dimensional scaffold, or 'fold'; and the total number of folds used by living things is likely to be as few as one to two thousand. The families of related proteins, which share folds and a common ancestry, are referred to as 'orthologs'. In cases where the set of related proteins is very large, they are referred to as superfamilies. Other groups of historically connected proteins arise from the duplication of ancestral genes and are referred to as 'paralogs'. The utilization of this narrow set of proteins constitutes another historical constraint. In November 2003 David Baker and his colleagues at the Howard Hughes Medical Institute showed that completely new protein folds, unlike any found in nature, could be designed and constructed from first principles. The introduction of these into synthetic organisms could confer completely new properties on them.

Most of nature's bias towards certain families and superfamilies is likely to result from their adaptive value, but it is possible that favourable families have been lost by chance or may, for exclusively historical reasons, simply not have been discovered. The current set may have been fixed because they were the first to be discovered, rather than because they were necessarily the best. Although only a tiny fraction of it is used in nature, the size of the potential protein universe is immense. For example there are 20^{100} (10^{130}) possible protein sequences for proteins that are 100 amino acids long and which have 20 amino acids available at any particular position in their polypeptide chain – a number bigger than the number of atoms in the universe. It has been estimated that fewer than 10^{50} different protein molecules could have been made during life's history on Earth, so an exhaustive search of even this tiny region of protein sequence space could not have been possible, even in principle. Artificial methods, however, could be used to identify new proteins.

Many types of constraints are fundamental and insurmountable; historical constraints, on the other hand, could be circumvented by constructing synthetic organisms from first principles. In a post-apocalyptic world consist-

7. THE LIMITS OF POSSIBILITY

ing only of gorillas, a skilful gorilla geneticist should in principle be able to reconstruct a lobster. Whether the environment would be sufficient to sustain synthetic lobsters, however, is another question. There are also creatures that we know once existed, such as dodos, whose reconstruction must therefore be possible. The existence of bizarre creatures such as the duck-billed platypus, the Madagascan giraffe-necked weevil, the pygmy hippo, seadragons and the pot-bellied seahorse shows that the realities of biological possibility are more complex than anything our imagination can conjure up. In his *Book of Imaginary Beings*, Jorge Luis Borges wrote that 'the zoology of dreams is far poorer than the zoology of the Maker'. There must be countless logically possible organisms that have never existed. Fairies may fall into this category. The inventory of organisms 'discovered' by history and evolution – of which around 1.7 million are in existence – represents only a tiny subset of life's possibility. The majority of living species, moreover, are insects. So how is the synthetic biologist of the future to proceed? Is there some method by which the complete inventory of biological possibility can be defined? Put simply, are living things computable? Given the necessary information about a potential creature, is there some way of determining whether it has a chance of being built and surviving? The question may be reformulated like this: what is the minimal information specification needed to fully describe a living thing?

The extent of the information needed for a full description of an organism depends on its complexity. At the simplest level, a living thing is specified by its genomic DNA sequence. In the case of the very simplest organisms – viruses – this is the only dimension of information necessary. But a virus is not a living thing: it is not free living and depends on the metabolic machinery of cells to replicate itself. For the sake of argument, however, let us imagine that all organisms could be fully described by their DNA sequence alone. One might ask whether there are any characteristic features within the genome of a creature – a giraffe, say – that would enable an alien from another planet to predict its appearance and behaviour. Could the alien also infer the appearance of a pelican or a python? Eventually the genomes of all known living things will be sequenced; at that point, or before, computer programs could be devised to search genome databases for features that indicate the type of creature each genome may encode.

At the most rudimentary level, the small genome of a bacterium would indicate that it could not encode anything as complex as a lion or a hippopotamus. The sequence of a human would similarly suggest that it is unlikely to compute anything as simple as a bacterium. Given genome databases of sufficient size, it should be possible to make more specific predictions. It would be surprising, for example, if the genomes of insects did not have features that distinguished them from fish. It is less clear whether it would be possible to distinguish the genome of a spider from a stick insect. But a DNA

7. THE LIMITS OF POSSIBILITY

specification may turn out to be sufficient to compute the morphology and basic behaviour of most creatures. If this were the case, then the same method could be used to search for creatures that have never previously existed or which are now extinct.

But how might this be achieved? Before addressing this, let us consider the analogy of a library. If a computer were programmed to produce random combinations of the 26 letters of the alphabet and punctuation marks to produce word strings anywhere from one to an infinite number of letters long, and the results printed out, an immense pile of books would be generated. The majority of these would be nonsensical in any language. But if a machine that printed infinitely fast and had an infinite supply of ink and paper and a storeroom of infinite capacity were used, it would be possible to generate a library so large and complete that scattered among the sea of random word strings would be the complete works of Shakespeare, Milton, Tolstoy and every piece of prose that has ever been – and ever could be – written. This is because the library would contain every possible combination of words of every possible length (limited only by time, storage space and paper) and would represent a universal library of all possible books.

The problem, of course, would be to locate islands of meaning in the sea of nonsense, since each meaningful book would be surrounded by irrelevancy. By analogy, it is possible to envisage a universal library called DNA space. This contains the complete collection of every possible DNA sequence. By definition, DNA space must contain the genomes of all known living things, since they are encoded by sequences of DNA. This immense space must also contain the genomes of all extinct creatures and those of all possible DNA-encoded creatures. The majority of these will never have existed. There has not been enough time or space to realize them. DNA space might include the flamingosceros, a hypothetical cross between a flamingo and a rhinoceros, and a kangapelican, a cross between a kangaroo and a pelican. A great many of these potential creatures will be logically flawed and unrealizable. There would be fish without gills, bison the size of ants and caterpillars as big as the Post Office Tower.

But the mathematical reality of all these creatures – both possible and impossible – should not be in question. One has only to think of dodos: if they had never existed, it would be hard to predict that dodo-like birds were a possibility. As it is, we know not only that the existence of dodo-like birds is possible but that they once lived quite happily in Mauritius. It might even be possible to rescue enough DNA from old flesh to rebuild one. Located in some obscure corner of DNA space we might – somewhere in a sea of false fairies that lacked wings, or were inordinately kind or oversized – find the perfect fairy genome we were looking for. One thing is for sure: the DNA specifications for dodos, and a host of other strange and never-realized creatures, have a timeless mathematical existence, irrespective of whether their real flesh and

7. THE LIMITS OF POSSIBILITY

blood manifestations happen to exist at any given time in history.

If DNA sequences alone turn out to be sufficient to compute the morphology and behaviour of the potential organism encoded by each possible genome in DNA space, the problem is greatly simplified. One would, however, expect that the computer program would have to be sensitive not just to the presence or absence of genes and their mutations, but also to epigenetic factors and non-coding DNA associated with each gene. An appropriate computer program could scan this information space genome by genome, and determine whether any given genome is nonsensical or encodes a potential living thing. Once an interesting genome had been located, the scientist's task would be to synthesize it to see whether it could be used to construct a living thing – thereby confirming or refuting the prediction. It is not quite this simple, however, since in many cases an appropriate egg would be needed for development, because genomes have to be kick-started into life. Other dimensions of information include the environment available to the organism. A panda without bamboo would be useless, since they eat nothing else. If the entire subset of DNA space that corresponds to the size range of known genomes were explored by computers, it should be possible to construct a map of the space. This huge multi-dimensional chart would mark the positions of all potential organisms. But the designation of morphologies would have to be provisional – until such a time as definitive proof of their shape and biological plausibility was provided by synthesizing the predicted organism and testing it in the real world.

It is possible that the information in DNA sequences is sufficient only to compute caricatures of organisms, and that additional tiers of information will be necessary to improve their predictive value. These missing dimensions of information include epigenetic information as manifested in the methylation patterns of DNA sequences. It may also include information about the egg into which a synthetic genome would need to be transplanted in order to kick-start it into life. Information obtained during development and by learning may also be important. In the most complex organisms, much of the behavioural information critical for survival is stored extragenetically in brains, or, in the case of humans, in cultural artefacts such as libraries and human memories. The behaviour of humans, more than any other animal, is the product of two analogous but distinct inheritance systems: genetic and cultural. Since the details of culturally acquired information have no DNA representation, DNA space cannot help reconstruct it. It might be possible to reconstruct a physical facsimile of an extinct dodo, with the morphology of the synthetic organism exactly resembling the real thing, but it is unlikely to be behaviourally equivalent to historical dodos. Their social and physical environments without doubt influenced their behaviour. The dodo's cry, for instance, is likely to have been transmitted culturally from generation to generation. It will therefore always be impossible to know exactly what a dodo sounded like: its cry is lost for ever. Synthetic dodos may look like the real thing, but the synthetic and ahistorical manner of their construction makes them different from the inhabitants of Mauritius that greeted sailors in the sixteenth century.

If we created a Noah's Ark containing DNA from all species, we should also record their cries and behaviour, so that if they were to become extinct, reconstructed versions could relearn some of their lost cultural heritage from videos and sound recordings. Socially acquired information is especially relevant to complex organisms. Wild Coho salmon are increasingly rare, though large numbers are raised in hatcheries for commercial harvesting. Conservationists point out that despite the abundant numbers of hatchery salmon, they display the stereotyped and impoverished behaviour found in zoo animals and behave differently from wild Coho. Differences in cultural attributes are apparent in the chimps inhabiting the Tai forest in the Ivory Coast. Tai chimps have invented a nutcracker: they position a nut on a stone anvil and bash it with a rock. Genetically near-identical chimps in the Gombe National Park in Tanzania have not learned this trick. But the Tanzanian chimps, unlike their colleagues in the Ivory Coast, have learned to use crumpled leaves to capture ants. Orang-utans also have a complex system of socially transmitted behaviour, as do elephants. In African elephants the oldest matriarch female is the repository of a group's social knowledge. The removal of older elephants by hunters can therefore have a profound effect. Whales also rely on extragenetic information, transmitting their vocalizations from generation to generation by cultural inheritance.

One intriguing question is whether evolution explores DNA space freely. The alternative is that DNA space is littered with 'attractor regions', consisting of genomes that draw processes of evolution into themselves, like black holes. If evolution were to be re-run on multiple occasions, attractor genomes of this sort would be expected to be discovered again and again. If so, humans and other species might be inevitable products of any evolutionary process, and the flora and fauna on Earth may be fairly common throughout the universe. But if the opposite were true, then if the history of life were wound back and run again from the very beginning, nothing like humans or the other usual suspects – mammals, insects, arthropods – would ever necessarily emerge again. So how robust are the grand patterns of evolution; are there only a few routes through DNA space, or many? Is the space so large that a process of life is statistically unlikely to pass through the same neck of the DNA woods again, or are there preferential mountain passes that channel processes of evolution to proceed from one region of DNA space to another few and far between?

Stephen Jay Gould has argued that the evolution of life on Earth has been characterized by a huge amount of historical contingency. According to Gould, we are the result of chance events rather than the inevitable products of an 'evolution machine'. Chance events can have a tremendous impact and are completely blind to their evolutionary consequences. If the dinosaurs had not been wiped out by the impact of a giant meteorite, life's history would have been very different: humans might never have existed. This is of course at odds with religious doctrines that see mankind as the key presence in the biological kingdom. It also has little in common with the theories of the palaeontologist Simon Conway Morris, who believes the constraints to be so imposing that humans are the likely end-point of any evolutionary process. The truth probably lies somewhere in between the two extremes. Evolution is both open-ended and subject to constraints, some of which can be circumvented, and which introduce a varying degree of bias into the manner and extent to which different regions of DNA space are explored.

The strongest evidence for developmental constraints and bias is found in the phenomenon of convergence, which describes the apparently independent evolution of similar characteristics in different species, or geographically isolated populations of the same species. A good example of convergence is found in the threespine sticklebacks of Canada and Iceland that live in lakes 5,700 kilometres apart from one another. Most threespine sticklebacks have a large spine on their pelvic fin, but some freshwater populations – including those in Paxton Lake in Canada and Lake Vifilsstadavatn in Iceland – have reduced or entirely lost this part of their skeleton. The loss or reduction of skeletal structures is as important in evolution as the emergence of new structures: the loss of appendages was crucial in enabling whales to adapt to new aquatic environments and for the emergence of snakes and eels; the reduction and loss of fingers and toes have been significant in the evolution of jumping, flying and running in a number of amphibians, reptiles and mammals. In April 2004 Michael Shapiro and his colleagues at the Stanford University School of Medicine showed that spine reduction in the threespined stickleback occurs by changing the expression of a single gene, *Pitx1*. In mutant sticklebacks, there is a localized decrease in *Pitx1* activity in cells within the appendages; in other body regions *Pitx1* is expressed as normal. The two isolated populations of sticklebacks in Canada and Iceland appeared to have converged independently on the same design and realized it using the same genetic mechanism. It was estimated that this major modification to the stickleback skeleton occurred over only 10,000 generations.

This extraordinary observation has two consequences. First, it shows that an apparently minor microevolutionary modification (an alteration to the regulation of a single gene) can result in significant macroevolutionary changes (the loss of a spine or a limb), proving that modifications to the software (mutations in non-coding DNA sequences) of developmental programs can produce fundamental morphological changes without altering the hardware (genes) themselves. Major modifications of this kind can be realized in natural populations within very short timeframes. These changes suggest a mechanism by which evolution could achieve some of its major transitions. When convergent evolution in two populations or species

7. THE LIMITS OF POSSIBILITY

occurs through the use of the same genetic mechanisms, the term 'parallel evolution' is used. Direct alterations to developmental genes themselves – seen, for example, in the hox gene *antennapedia*, which when mutated produces flies with legs sprouting from their heads – do not seem to provide fruitful routes for evolutionary change. This is probably because many of these genes have multiple effects on the developing body: it is one thing to lose the expression of the gene in an isolated compartment, but quite another to lose expression completely.

Second, it suggests that *Pitx1* and the non-coding software that controls its expression may represent a 'hotspot' for evolutionary change – a constraint on the manner and direction in which these organisms can alter their form. Why this is the case is still unclear. The fixation of mutations into DNA sequences is known to result from the way in which repair enzymes process damage to DNA; the frequency and nature of the mutations in any specific region of the genome is controlled by a number of factors, including the ordering of the nucleotides in the sequence. It is possible that hotspot regions either attract more mutations than other equivalent regions, or that mutations occurring at the standard rate are fixed more readily. Whatever the precise mechanism, mutations focused on developmental hotspots within key genes may be of great significance, perhaps in some cases driving evolutionary change in one direction rather than another. The fact that *Pitx1* is not expressed in the pelvic regions of limb-reduced fish in Scotland indicates that mutations in the *Pitx1* developmental control system may underlie the loss of limbs occurring in the convergent evolution of a host of different species, perhaps even whales.

Another example of how shared genetic mechanisms can underlie convergent or parallel evolution is seen in the fruit fly *Drosophila*. Some *Drosophila* species have independently evolved tiny bald patches. In August 2003 Elio Sucena and his colleagues at Princeton University showed that bald patches had evolved independently in three *Drosophila* subspecies, in each case resulting from the loss of a single gene, *shavenbaby*. The fact that the same genetic mechanism has been independently applied to achieve a common morphological end in different subspecies supports the idea that some developmental regulatory circuits preferentially accumulate evolutionary changes, and that this type of morphological parallelism may be relatively common. In this case the non-coding DNA software associated with the *shavenbaby* gene is a key regulatory point, an evolutionary hotspot that seems to constrain the scope of evolutionary possibility available for this particular characteristic of flies.

The mechanism that causes preferential accumulation of mutations in *shavenbaby* is a mystery, but the extent to which any particular genetic circuit – the hardware and its associated software – constitutes a developmental hotspot may depend on its connectivity: that is, on whether it integrates multiple inputs to generate a final output. Low connectivity circuits are more likely to fix mutations as the effects of the mutation are more localized. The pheno-

7. THE LIMITS OF POSSIBILITY

menon of parallel evolution is immensely important: it provides a way of integrating evolution with the genetic circuits that influence development, and illustrates the power of the concept of 'regulatory mutations' that do not affect the sequence of the gene itself, but have a profound influence on the program which determines how that gene is expressed in various body parts throughout an individual's lifetime. Major morphological innovations, then, can be achieved not only by the evolution of new genes, but also, more simply, by rewiring developmental circuits and putting the same set of genes to use in different ways.

But the modification of regulatory circuits is not the only mechanism by which parallel evolution proceeds. Mutations in the genes themselves are also sometimes implicated. The convergence of different species of birds on similar plumage patterns is widespread and provides a good example of this. There are 26 different species of birds – including the black-capped chickadee and black-capped pygmy tyrant – whose head plumage is distinct enough to give them a 'black cap' prefix. Similarly, there are 41 black-throated species: eight that have blue caps, nine that are orange-breasted and 29 that are red-billed. Nicholas Mundy and his colleagues at Cambridge University studied a plumage pattern that has evolved independently in two distantly related species of arctic birds: the lesser snow goose *Anser caerulescens* and Arctic skua *Stercorarius parasiticus*. Both species exist in either a dark ('blue' snow geese and 'dark' skuas) or a light form ('white' snow geese and 'pale' skuas). These different plumage versions influence mate choice. Mundy showed in March 2004 that convergence on the same characteristic results from a single mutation in the *melanocortin-1* (*MC1R*) gene.

The fact that the same point mutation underlies the independent evolution of the same characteristic in two distantly related species indicates that genes can form attractor regions in DNA space, just as regulatory sequences can, making it more probable that evolution will follow particular pathways. Some feature of the responsible region of *MC1R* presumably makes it hypermutable, and a hotspot for evolution. Such studies show that the evolution of new characteristics can be seen as the result of an intricate balance of environmental selection (for the 'fittest' version of a characteristic) and internal constraints (sometimes called 'developmental drive') that in some instances bias the intrinsic direction of evolution and make some outcomes more probable than others.

Such constraints, which channel evolutionary processes, are likely to operate at the level of micro-characteristics, as well as of macro-characteristics such as appendages and plumage. In November 2001 Jack Szostak, working at the Massachusetts General Hospital in Boston, showed that the hammerhead ribozyme – a unique type of RNA capable of functioning as an enzyme – has evolved independently on multiple occasions. The notion that the direction of evolutionary change can be determined by intrinsic factors and developmental dynamics contrasts with the historical thrust of Darwinism, in which evolution is driven exclusively by natural selection. The extent to which the directionality of evolution is bias-led remains to be seen. But the power of point mutations in genes and non-coding regulatory regions to reconfigure living things unpicks the mystery of speciation. It shows how the variety of species on Earth – in Darwin's words, the 'endless forms most beautiful and most wonderful' – were able to arise rapidly and by simple means. The raw power of genetics and evolution has been laid bare.

The overall pattern of life's unfolding is ultimately constrained at its origin. Until we understand the way in which life can be initiated from collections of inanimate molecules, it is hard to know how many routes lead out of lifelessness into the infinite expanses of heredity and DNA space. It may be that such routes are various and that each sets life off on its own particular journey. Or it may be that the number of routes is small, which would mean that however many times the program of evolution were run life would be stuck on a rigid and invariant railroad whose every twist and turn through DNA space is tightly controlled by intrinsic constraints. If this were the case there would – if the role of historical contingency were minimized – be a finite probability associated with the eventual appearance of all known creatures, including trilobites, cockatoos, porcupines, snails, zebrafish and humans. But however deeply affected by bias genomes turn out to be, the many dimensions of DNA space – and such tricks for leaping through it as the reshuffling caused by transposons – are likely to ensure that escape hatches or wormholes are available at every point, enabling processes of evolution to derail from the track set down by intrinsic constraints, free-wheeling onto a different track, one that is equally constraint-driven but which runs in a different direction. The intrinsic flexibility and plasticity of genetic processes ensures evolution's relentless procession through time and the twisting and buckling landscapes of DNA sequence space.

8

A manifesto for life

Some time in the late sixteenth century, Spanish travellers in central Peru encountered an old Indian man, possibly a former official of the Incan empire conquered by Francisco Pizarro in 1532. According to the account of Diego Avalos y Figueroa, it was clear that the man was trying to hide something; on searching him they found several bunches of cryptic knotted strings known locally as khipu, or quipu. When asked what information they contained, he replied that his khipu recorded everything that the Spanish conquistadors had done in the area, both good and evil. The leader of the Spanish party promptly burned the khipu and punished the man for possessing them. According to colonial accounts, khipu were kept by elite bureaucrats known as khipukamayuq or 'knot keepers', who knew how to parse the knots by running their fingers along them, as though reading Braille. So important were the knotted strings to this pre-colonial civilization that in 1542 the colonial governor of the region summoned local khipukamayuq to decipher their strings in order to help him assemble a history of the Inca people.

The Spanish considered the khipu to be dangerous, idolatrous objects – recording narratives, myths and other subversive information – and set about destroying as many of them as they could get their hands on. Only around six hundred survive today. But many of them were probably nothing more than mnemonic devices or accounting tools. The historian L. Leland Locke showed in 1923 that the one hundred or so khipu stored in the American Museum of Natural History in New York encoded nothing more seditious than the results of mathematical calculations. The Incas were long considered to be the only major Bronze Age civilization to lack written language. But contemporary khipu scholars now believe that though khipu may have begun as accounting devices, they later evolved into a three-dimensional binary narrative code. At least a fifth of surviving khipu appear to be non-numerical. This writing system – if indeed that was what it was – seems to be unique, since all other such systems use calligraphic implements to inscribe characters onto two-dimensional surfaces.

In 1997 William J. Conklin, an expert in textile designs, suggested that the knots might represent only part of a khipu's complexity. He conjectured that up to 90% of the information was present in strings before any knots had been tied. According to Conklin, the narrative within each khipu text was encoded in a set of binary alternatives. These included the type of material (cotton or wool) from which they were made, the spin and ply direction (the threads slant in either an 'S' or a 'Z' direction), the direction of the knots that attach hanging subsidiary strings to primary strings, and the direction of slant within

8. A MANIFESTO FOR LIFE

each knot. These complex string lexicons are one of the few examples of semasiographic writing: texts that are not representations of spoken language. Unfortunately, the khipu code has not yet been convincingly deciphered. It is hoped that a khipu will eventually be found with a corresponding Spanish translation, to enable the khipu code to be cracked and the voices of Inca ghosts to speak through these ancient knotted strings.

The digital way in which cultural information is recorded in khipu corresponds to the way information is encoded in DNA using combinations of the four nucleotide bases. Both generative systems are 'open-ended': they allow infinite use to be made of finite means and have a prodigious capacity for information storage. Different combinations of DNA sequences can represent the genomes of all known species. Analogously, the digital knot code of khipu can represent most types of cultural information. A complete collection of the khipu transcribed before their destruction by the Spaniards would have given a detailed description of Inca life, including their history, practical knowledge and customs. Khipu constituted a sort of 'cultural genome' for the Incas, encoding aspects of their existence that were not captured by DNA. Unlike genes, which record only the tried and tested wisdom of ancient history, cultural databases hold information of a more contemporary nature. The permanent loss of cultural records through the destruction of artefacts is in some ways analogous to the extinction of a biological species, since these records represent an important dimension of the nature of complex organisms. Stripped of their cultural heritage, modern humans would revert to the behaviour and customs of our *Homo sapiens* ancestors who inhabited the grasslands of Africa. Despite their genetic similarity, these two states of human existence are so distinct that they behave almost as if they were separate species: their cultural differences are far more significant than any anatomical differences. Humans have evolved an extraordinary capacity for cultural diversity. The fact that this is sustained by an underlying genetic homogeneity illustrates the intrinsic plasticity of culture and its relative freedom from genetic constraints.

In the three billion or so years before culture evolved, the heredity systems of living things were based exclusively on DNA. But with the emergence of culture – the inter-generational transmission of socially acquired information by imitation and, in the case of humans, through the production of artefacts – a new parallel system of heredity was introduced. For the first time in history, part of the information of a species could be stored in a medium other than DNA. In most contemporary animals this extragenetic record is housed in the patterns of connections between brain cells. In humans the media of storage can also be non-biological. The most powerful storage systems are abstract and generative. They include the symbolic grammars of standard writing, semasiographic writing of the type seen in khipu, spoken language, the Arabic numerals used in mathematics, computer programs, labanotation (a system for coding choreography)

8. A MANIFESTO FOR LIFE

and musical notation. Non-generative media – such as cathedrals, paintings and tools – are equally powerful but less flexible vehicles of cultural expression. They are analogue rather than digital, unlike DNA and language, which encode information in an abstract combinatorial form that has no structural correspondence to the entity being described. In analogue systems, the object itself stands as a direct record of the information without employing mechanisms for coding and decoding, the difference being akin to that between an old gramophone record and a compact disc.

Extra-genetic and extra-biological systems of information transfer and inheritance complement DNA-based systems of heredity; their emergence has been hugely significant in human evolution. They vastly increase the amount of information available to a species and allow each generation to benefit from the experience of its predecessors. Without such mechanisms, every aspect of human knowledge – the use of fire, the wheel – would have to be rediscovered by each new generation. The relationship between genes and culture is difficult to determine, since socially transmitted information blurs the distinction between what is innate and what is acquired. But though culture seems independent of their influence, genes must to some extent constrain the broad types of possible cultures.

Individual elements that sustain aspects of a given culture, such as an ability to appreciate music, are in some cases under strong genetic control. Che Guevara suffered from congenital amusia, a type of near-total tone-deafness that turns music into noise. Up to 5% of some populations suffer from this syndrome; Theodore Roosevelt and Ulysses S. Grant were both affected. Isabelle Peretz, working at the University of Montreal in Canada, studied a patient called Monica, who was unable to recognize or sing familiar songs such as *Frère Jacques*, despite having normal mental functioning in other respects. In biological systems, mating barriers restrict the flow of genetic information between species; humans have developed analogous ways of restricting the transmission of socially acquired information. Human cultural barriers to information transfer include social class, secrecy laws, patents, copyrights and other intellectual property rights. The importance of culturally transmitted information is neatly illustrated by the fact that around 40% of the inmates of the Paris Bastille in 1759 had been sentenced for offences connected to the book trade.

Before the human genome was sequenced, it was assumed that very large numbers of genes are used in the construction and operation of humans: this seemed a sufficient explanation for their complexity, including their unique capacity for extended culture. The discovery that humans have only around 24,500 protein-encoding genes, rather than the predicted 100,000, was for this reason a great surprise. Early estimates suggested that we share around 99% of our DNA with chimpanzees and slightly less with gorillas. Our apparent genetic similarity to chimpanzees is so great that Jared Diamond has referred to humans as the third species of chimpanzee. It seems that

8. A MANIFESTO FOR LIFE

only a tiny portion of the DNA in our genomes is responsible for making us human. Despite being almost genetically identical to chimps, however, we do not look or act like them. So why are we so different? How has it come about that we are hairless, walk upright, have great manual dexterity, write novels, compose symphonies, are able to converse over a glass of wine, recite poems, generate civilizations, send spacecraft to distant planets and agonize about the meaning of existence – when all other creatures live mundanely from day to day? What is it that makes us human?

The emergence of all human characteristics – from cellular structures to consciousness and the capacity for extended culture – has a rational basis in the hardware and software modifications to our genomes. These modifications have resulted in the creation of new genes, the alteration and deletion of old ones, the generation of new patterns of gene activity and the horizontal gain of genes from bacteria and viruses (around two hundred human genes are thought to have been acquired from bacteria in this way). Irrational explanations of human uniqueness, and in particular of our mental capacity, invoke metaphysical forces – sometimes referred to as the soul or spirit. The dualistic perspective, which envisages humans as a *machina carnis* or 'body machine' consisting of a material body coupled transiently to an immaterial soul, was promulgated by the philosopher René Descartes. Modern science denies such agencies, insisting that consciousness and other mental phenomena are explained by brain function. Although the Catholic Church now accepts the principle of evolution, it excludes the human soul from it. It may be that one of the things that make us different from other primates is our possession of genes that make us believe in concepts like the soul. A comparison between the genomes of believers and non-believers might help identify some of these. The developmental psychologist Paul Bloom has argued that our tendency to perceive ourselves as dualistic composites of material and immaterial agencies is hard-wired from birth. Our tendency to believe in an immaterial soul is, it seems, a consequence of the way in which our brains and cognitive systems have been assembled by evolution – making us natural-born dualists. In September 2002 Olaf Blanke, at Geneva University, discovered that out-of-body experiences – sometimes claimed as proof that humans have a soul – can be induced by stimulating the brain's right angular gyrus. The occurrence of metaphysical beliefs does not mean that the notion of a soul has any meaning or existence independent of our brain's constructions. The fact that the mind is a product of brain function does not similarly necessarily destroy metaphysics, but it does make the conventional exposition of metaphysics less plausible.

If all human characteristics, including consciousness, can be understood without invoking explanations outside biology, it seems inevitable that we will have to resign ourselves to the unpalatable fact that we are nothing more than machines. We may be more complex than toasters or dishwashers and more sophisticated than dandelions, but we are machines nevertheless. That this troubles us is itself a construction of our brains; one day such irrational tendencies might be removed by adjusting the relevant brain circuitry. Any definitive demonstration that humans lack an immaterial soul should not cause undue theological concern, however, since there are a great many things in the universe that biology cannot explain. We do not have to resign ourselves to the fact that the human genome is the secular equivalent of the soul. Even if all aspects of our humanity can be explained by the mechanical operations of the body, it is possible that there is a dimension to human existence that has no role in routine corporeal phenomena such as consciousness and that is not generated by gene function. If such phenomena do exist, they must arise from material extracorporeal agencies and parasitize human brains in order to coexist with the host's body without being generated by it.

Once the genes and programs that make us human have been identified, we might choose to transfer them into other species in order to humanize them. By doing so we might be able to make monkeys, pelicans or porcupines think; it is perhaps less likely that we could perform the same transformations on centipedes or marigolds. These differences would help identify the minimal architectures – probably including large brains – necessary for sustaining complex intelligence. We should remember that a full description of humans must include information acquired during development, by learning and cultural transmission, but the genetic description is fundamental: it provides the basic sketch of our humanity.

The simplest way of identifying the genes that make us human is to compare the sequences of our genomes with those of other creatures. The more similar a species is to us, the more valuable such comparisons are likely to be. The differences between humans and chimps, say, would point to candidate human genes possibly responsible for language, culture, imagination, emotions and thought. The sequencing of the human genome is in this sense not an end in itself, but rather the beginning of a new approach to biology in which comparisons between the genomes of different species illuminate the mechanisms that make each unique.

When completed, the chimpanzee genome whose draft was assembled in December 2003 by Eric Lander and colleagues at Washington University is likely to yield much useful information about the genes that make us human. The complete and detailed sequencing of chimpanzee chromosome 22 in May 2004 by an international consortium is the first step in this direction, leading to a direct comparison with its human counterpart, chromosome 21. 83% of the corresponding human genes had alterations that result in modified proteins. Some of these were structurally minor, changing only one or two amino acids but, unexpectedly, 68,000 insertions and deletions (indels) – in which large or small regions of DNA had been either gained or lost – were found spread across both the coding and non-coding DNA. The larger indels appear to be a result of the action of transposable elements. 99% of the indels were less than 30 nucleotides long, but of the larger ones some extended to 54,000 nucleotides.

This suggests that indels were one of the major factors responsible for remodelling the chimpanzee genome into a human one. Of the estimated 284 protein-encoding genes in chimpanzee chromosome 22, 87 had mutations that altered at least one splice site, indicating that some of the differences between chimps and humans result from the production of alternative forms of the same genes. A study of this kind is a start, but it cannot establish which is primary, the loss of DNA or its gain. For this the study has to be broadened to include other types of primates, such as gorillas and orang-utans. This has been done to a very limited extent, making it possible to show that one particular sequence, at least, was gained by humans rather than lost by chimpanzees. But more extensive studies encompassing many primate species are needed to clarify these issues.

Michele Cargill and her colleagues at Celera Diagnostics adopted a different approach. They compared the sequences of known chimpanzee genes with those of humans and mice, focusing on the 7,600 genes that were shared by all three species, out of a starting pool of 200,000. These were examined in order to determine which had undergone accelerated evolution through natural selection. 1,547 human and 1,534 chimp genes of this kind were identified, indicating that the mutated versions had distinct survival advantages. Analysis showed that accelerated evolution was principally restricted to certain functional classes of genes. In humans the categories showing the greatest acceleration included those involved with the sense of smell, hearing, developmental processes and amino acid metabolism. In chimps, the categories included those involved in DNA repair, skeletal development and cell structure. It is likely that the genes in humans that have been subjected to accelerated evolution may also be involved in disease susceptibilities – so studies of this kind may have the added advantage of throwing light on certain diseases by identifying susceptibility genes.

Nothing is more quintessentially human than speech and language, which are likely to have been prerequisites for the development of extended culture. Bees dance, lions roar, insects buzz, birds sing, chimpanzees grunt, but animal communication lacks the open-ended power of human language. Language has three fundamental elements: phonological (sound production), syntactic (the sequencing of words to produce grammatical structures) and semantic (meaning). Each may be associated with a corresponding brain region that performs the necessary computations. The production and reception of language also requires finely tuned sensory and motor systems. In *Syntactic Structures* (1957), Noam Chomsky argued that the study of language is related to the study of mind, since to describe language is to describe a central feature of the mind. He suggested that humans are born with brain structures that encode an innate universal grammar which

8. A MANIFESTO FOR LIFE

enables them to acquire the details and shared logical structures of any human language. The implication is that genes involved in the evolution of language are likely to be linked to the evolution of the mind.

There is plenty of anecdotal evidence that animals have rudimentary minds. Frans de Waal, for example, described a troublemaking female chimpanzee living in a colony at the Yerkes Regional Primate Center in Atlanta, Georgia. When visitors arrived, she would take a mouthful of water, casually mingle with the rest of the monkeys, and wait with closed lips until the visitors drew near. Choosing her moment with care, she would then spray them and shriek with laughter. Animals grunt, gesticulate and display behaviour that suggests an underlying rudimentary mental representation. But they cannot speak. Speech is arguably the single most important factor differentiating between humans and other creatures. If there were a gene for speech, it would be a good candidate for a gene that makes us human.

The capacity of human language to generate a limitless range of meaningful expressions from a finite set of simple elements distinguishes it from all other animal communication systems. The rule systems that generate infinite sets of outputs ('grammars') vary in their generative power. A key aspect of human language is its capacity for recursion. Recursion makes it possible for the words in a sentence to be widely separated, but nevertheless dependent on one another; it enables language to use abstract hierarchical structures to generate complex constructions and concepts. An example of recursion is found in 'if-then' sentences. In the sentence 'If you do not book your plane ticket early, then you will not be able to go on holiday', 'if' and 'then' are dependent on one another, even though they are separated by an arbitrary number of words. In January 2004 Marc D. Hauser and Tecumseh Fitch at Harvard University showed that although tamarin monkeys were able to master a weak, non-recursive grammar with local organizational principles, they were unable to learn more sophisticated recursive grammars.

The over-reliance on superficial aspects of stimuli prevents tamarins from perceiving more abstract relations present in the signal. Recursive grammars, which give human languages their limitless possibilities for expression, enable the generation of structures like phrases and sentences that go beyond the simple word level (or single call level in animals). Linguistic syntax involves the permutation of such structures to generate higher-order meaning. Since language is such an important human attribute, it would appear that a defining moment in human evolution occurred when the computational constraints on syntactic processing that characterize the communication systems of non-human primates were transcended. As with all other fundamental human competences, the ability for higher-level syntactical processing is likely to have a genetic basis.

A large British family, known as KE, that has suffered for several generations from a severe inherited speech articulation and language disorder has helped identify one of the best candidates yet for such a language or syntax gene. Around half the family members have speech so deranged that it is barely intelligible – this in spite of their having adequate intelligence and no neurological or sensory deficits. The abnormal speech is the result of a combination of severe articulation difficulties and disorders of fine motor control of the mouth and tongue, coupled with linguistic and grammatical impairments. The *FOXP2* gene, discovered by Anthony Monaco and Faraneh Vargha-Khadem at Oxford University in the autumn of 2001, is now known to underlie this disorder. *FOXP2* is unlikely to be the only gene involved in language: its discovery is a bit like finding one part of a car that initially looks useful but turns out to require related machinery.

Affected individuals have a single change in *FOXP2* that disrupts its functions, one of which is to regulate the early development of brain structures responsible for speech and language. The mutation changes a single guanine nucleotide in the gene to an adenine: the letter G in the DNA sequence is replaced with an A. This results in the substitution of a histidine amino acid in the *FOXP2* protein for an arginine. Affected members of the KE family have only one abnormal copy of the gene, indicating that normal development of the regions of the brain responsible for these aspects of speech and language cannot occur if only half the normal amount of active *FOXP2* protein is present. *FOXP2* encodes a transcription factor, a class of proteins that regulate gene expression. The point mutation occurs in a region of key functional importance in the protein known as the 'forkhead domain'. A tiny change to a single gene can thus disrupt a fundamental human characteristic. Just two amino acids distinguish the human *FOXP2* protein from the chimpanzee version. The human version of the gene has been shown to date back no more than 200,000 years, to the time when anatomically modern humans first emerged. Given that language is a prerequisite for extended culture, were the mutations in *FOXP2* necessary for the evolution of humans? Might the acquisition of these changes have helped set our distant ancestors on a trajectory that led to humanity?

Advanced mental abilities and a capacity for finely-tuned manual dexterity enabled humans to dominate the rest of the animal kingdom. These were made possible by the disproportionate expansion of the cerebral cortex, a thin sheet of neurones that constitutes the outermost layer of the brain. The thickness of the cerebral cortex has remained fairly constant over the course of evolution, but its surface area has expanded enormously. The surface area of the human cerebral cortex is around one thousand times larger than that of a mouse, though it is only twice as thick. The cerebral cortex of mice is entirely smooth; that of humans and other primates is extensively wrinkled and creased, giving it a 'gyrencephalic' shape. Its convoluted appearance is a purely physical phenomenon, arising from the difficulty of stuffing an enlarged volume of brain into the close confines of the skull. This results in the brain buckling and folding, like the rucks seen in a

8. A MANIFESTO FOR LIFE

carpet when it is squeezed into a room that is too small for it.

Genes that increase the size of the cerebral cortex, if any exist, might be the genes that made us human and facilitated the rapid cultural evolution and geographical expansion that began around 50,000 years ago. This is because cortical surface area expansion in humans greatly exceeds that seen in other primates. Working at the Brigham and Women's Hospital in Boston, Anjen Chen and Christopher Walsh speculated that a protein called beta-catenin functions to influence cell numbers and cell fate decisions during brain development. In July 2002 they showed that if the amount of beta-catenin protein in developing mouse brains was artificially increased, it resulted in mice with grossly enlarged brains, with an enlarged cortical surface area of normal thickness. This artificially induced growth was so pronounced that the normally smooth mouse brains developed convoluted folds reminiscent of the gyri (hills) and sulci (valleys) seen in human brains. It seems that the apparently difficult transition from small to large brains can be achieved through the implementation of a simple genetic trick.

A study published in March 2004 by Xianhua Piao at Harvard Medical School indicates that single gene mutations can also influence regional development of the cerebral cortex. This is significant, since the cerebral cortex is not homogenous but is divided into dozens of well-defined areas, each with a specialized function. In humans, the frontal lobes – essential for social behaviour, cognition, language, problem solving and motor function – are especially prominent. The differential expansion of this brain region was clearly crucial to human evolution. Piao studied patients with a genetic disorder of cerebral cortex development known as bilateral frontoparietal polymicrogyria (BFPP) and showed that these patients had mutations in a gene called *GPR56*.

This random experiment, generated by nature, provides an important insight into how we became human. Examination of the brains of affected individuals revealed that their frontal lobes had a severely abnormal architecture, as did their parietal and posterior cortical regions, though to a lesser extent. At the microscopic level, the cortex in affected areas was noticeably thinner than normal, with four rather than the usual six layers. In a parody of normal sulci and gyri, the cortex of BFPP patients is thrown into an innumerable number of tiny convolutions. These gross structural abnormalities result in a host of malfunctions – including mental retardation, language impairment, gait abnormalities and seizures. A range of similar syndromes selectively target other regions, such as the occipital cortex, suggesting that the local microanatomy of the cortex is controlled by a host of different genes: the various cortical areas have therefore been able to evolve semi-independently, allowing for the evolution of specialized cognitive abilities. It is likely that the expanded frontal cortex of humans is the result of the acquisition of mutations in *GPR56* that are absent in other primates.

The dramatic expansion of the brain in human primates

8. A MANIFESTO FOR LIFE

could not have occurred in isolation. Although it is still tightly packed, the human skull has increased in size considerably in order to accommodate its mass of brain tissue – suggesting that any mutations that resulted in an increased brain size must have been closely linked to mutations that enabled the skull to enlarge. Are there genes capable of achieving this complex feat, giving the human skull both its characteristic shape and the increased volume necessary for brain expansion? The fossil record supplies some insights here: it turns out that the emergence of modern humans was accompanied by a connected web of morphological innovations. These innovations helped distinguish early humans from the ape-like hominid species that preceded them. Two and a half million years ago, all species of hominid had massive jaw muscles and prominent bony protuberances on which to anchor them. But only *Homo erectus* came to look human. Its facial features and jawline gradually softened, and its enormous masticatory muscles shrank to a modest sliver. Significantly, these changes were correlated with an increase in cranial capacity and brain size. Is it possible that the reduction in the bulk of the jaw muscles was responsible for cranial expansion?

It is hard to believe that something as simple and apparently inconsequential as a change in the bulk of the muscles involved in chewing could influence the shape of a rigid structure like the human skull. It is even more extraordinary that this could have changed the course of human evolution. But studies have shown that alterations to the size of muscles can have dramatic effects on the anatomy of the bones to which they attach. Computer simulations show that changes in muscle bulk can radically alter the growth pattern of the craniofacial skeleton. This is probably a result of the physical stresses that contracting masticatory muscles exert on the bones of the brain cases, which, during their growth phase, are very pliable. Movement and growth of the bony plates from which the skull is made are impaired by excessive muscle bulk. This suggests that factors that decrease the bulk of the masticatory muscles, and therefore the force of their contraction, could remodel the cranium. The great bulk of our hominid ancestors' chewing muscles, essential for their meaty diets, may have prevented them from evolving into humans. But what could have led the jaw muscles of our *Homo erectus* ancestors to shrink so dramatically?

One of the components of the microscopic units of skeletal muscle, known as sarcomeres, are the myosin heavy chain (*MYH*) proteins. These come in a number of different types, each specialized for different rates of muscle contraction. The inactivation of individual *MYH* genes dramatically reduces the size of the muscles in which they function. In March 2004 Hansell Stedman and his colleagues at the University of Pennsylvania showed that the *MYH16* gene, which encodes the version of the *MYH* protein specific to the jaw muscles in humans and other primates, is mutated in humans but not in monkeys. This mutation truncates the human *MYH16* protein, resulting in an ineffective protein roughly two-thirds of its normal size – and in a greatly reduced jaw muscle bulk. All non-human primates lack this mutation and produce a full-length *MYH16* protein. Stedman showed that this mutation probably pre-dated the emergence of the modern cranial architecture around 2.4 million years ago. This is a further indication of its likely importance in the evolution of modern humans.

But the emergence of human morphology and advanced cognition was not only dependent on modifications to the genetic hardware of early hominids, or on the evolution of new genes. Extensive reprogramming of genetic software was also necessary. Evidence for this has been provided in a beautiful experiment carried out by Wolfgang Enard and Svante Paabo at the Max Planck Institute in Berlin. They collected samples from the brain, liver and blood of humans, chimps, macaques and orang-utans and isolated mRNA from each tissue. This was passed over a 'gene chip' that contained tags for around 12,000 human genes. Gene chip technology enables the activity patterns of genes – the gene expression programs – to be observed directly, by attaching thousands of pieces of DNA onto a glass surface. Each gene is represented by a dot of DNA on the grid. If the activity of the gene increases, mRNA from the sample will bind to the chip and the signal at the corresponding position will similarly increase. If, on the other hand, the level of gene expression goes down, or if the gene is switched off, the signal will attenuate or disappear. Gene chips are like a new kind of microscope, allowing the activity of genes to be observed. Since human genes are very similar to those of other primates, human gene arrays can also be used to determine the gene expression patterns in monkeys.

Although only very few differences were found between the various primate species when their blood and liver gene expression programs were compared, significant differences were seen between the programs in chimp and human brains; the expression patterns in chimp brains were fairly similar to those seen in other monkeys. The fact that the programming of the human brain differs so significantly from that of other primates indicates that the human brain was subject to accelerated evolution in a way that other organs were not. This study provides the first clear evidence that the evolution of humans involved unusually rapid changes in gene expression. The reprogramming could have occurred at any time during the course of hominid evolution and helps explain how humans and chimpanzees can have such strikingly different morphologies and cognitive capacities, despite having large numbers of genes in common.

The genes responsible for helping to make us human influence every aspect of brain function and behaviour, from our ability to learn and remember to our moods, emotions, sociality, perception and cognition. In rare cases single genes have been implemented, but evidence suggests that in most cases collections of interacting genes are responsible, some of which are environmentally sensitive to experience in early development. The importance of experience in helping to influence genetic programming is demonstrated by the fact that monkeys raised with inanimate objects as surrogate mothers are found to suffer from social deficits later in life. A lack of early sensory experience also has an adverse effect on the developing visual cortex of the brain, whose microanatomy is contingent on experience-dependent plasticity. The influence of genes can be pervasive and unexpected, affecting aspects of our behaviour that intuitively appear to be as far removed from genetic influence as is possible. Onur Gunturkun studied kissing in public places – railway stations, beaches, parks and international airports – and showed in February 2003 that around 65% of couples turned their heads to the right when they made their first embrace. The basis of this behavioural asymmetry, which first occurs in childhood, has not yet been established, but it is likely to be genetically determined.

The human brain is a complex network of around one hundred billion nerve cells and trillions of specialized points of contact, known as synapses. Synapses convert the electrical signals in nerve cells into chemical signals, then stimulate a new electrical signal in the nerve cell at their other end – a process that allows nerve cells to communicate with one another. The chemicals released in synapses are known as neurotransmitters. The human brain is a unique biological machine – evolved by trial and error over 1.2 billion years – responsible for all mankind's great cultural achievements, from *The Marriage of Figaro* to spaceships, from the pyramids to the Parthenon. There is in principle no reason why non-biological or artificial biologically based machines should not generate minds, or – given the continuity of all living things – why animals should not be attributed with simple minds. Even plants demonstrate intelligent behaviour. A growing shoot can sense its nearest competitive neighbours (using near infrared light), predict the consequences of their activity and, if necessary, take evasive action. All of this is computed without a brain. So brains are not the only type of biological architectures capable of generating intelligence. But minds, unlike computers, do not just manipulate symbols: they also attach meaning to them. It would take a very special type of artificial computer to generate a mind, perhaps one based upon biological principles.

Situations where the brain malfunctions – in patients with schizophrenia or depression, for example – can provide insights into the genes that help shape the mental world, including our moods and thoughts. In schizophrenia, mental activity is profoundly disrupted: patients exhibit psychotic symptoms, including delusions and auditory hallucinations. The disease is also associated with a wide range of cognitive deficits, including flattened emotions, disordered thinking, social withdrawal, attention deficits, and problems with planning, problem solving and short-term memory. It is now thought that in schizophrenia the cognitive disorders are the primary problem, and that psychotic symptoms are secondary consequences. The mechanism for schizophrenia is still unknown, but the fact that between 20% and 50% of the first-degree relatives of sufferers exhibit some of the disease's features indicates an underlying genetic component. One possibility is that multiple deficiencies in various cognitive functional modules accumulate until they reach a threshold, which in the right environmental circumstances produces schizophrenia. A gene called *COMT*, thought to be involved in the cognitive module that helps with planning and problem solving, seems to be associated with the disease. Relatives without schizophrenia may have deficiencies in some or several of these cognitive modules, but since they do not have the full house they do not develop psychotic symptoms.

The way in which some genes are sensitive to environmental influences is illustrated by a gene called *5-HTT*, one version of which appears to play a role in determining whether people get depressed in response to life stresses. *5-HTT* is a chemical transporter that fine-tunes the amount of the neurotransmitter serotonin in the brain's nerve cells and is a target for anti-depressive drugs. The gene comes in two versions, one with a long promoter and another with a short promoter. The short version produces only half as much *5-HTT* transporter protein as the long version. Animal studies showed that mice with two long versions of *5-HTT* coped with stress very badly: for instance, they exhibited more fearful reactions to loud sounds. In a spectacular examination of the effects of different forms of this gene in humans, Avshalom Capsi and his colleagues at King's College demonstrated in July 2003 that people with two short versions of the gene had twice the risk of developing a major depressive episode in response to stressful life events when compared to people with two long copies of the gene who had experienced similar stress levels. Having two copies of the short version of the *5-HTT* gene did not increase the risk of depression in patients who had not had stressful life events. It seems that stress is a trigger that unleashes the otherwise silent effects of this gene on information processing – demonstrating how a single genetic mutation can influence the way the brain interprets its environment and thereby affect behaviour.

Animal studies have helped identify some of the genes that influence behaviour: they suggest that humans might be influenced in similar ways. In September 1998 Laura Nelson at Boston University showed that nematode worms with a disrupted *flp-1* gene are unco-ordinated and hyperactive. An equally powerful gene controlling behaviour has been identified in mice. Usually mice sniff out the sex of a prospective partner before trying to mate, but if their *TRP2* gene is disrupted, the mice are unable to smell the sex of their partner and so by default always choose to mate rather than be aggressive. In July 1996 Jennifer Brown at Harvard Medical School showed that mice with an inactivated *fosB* gene refuse to nurture their offspring, ignoring them to such an extent that they die. More complex behaviours can also be influenced by genes. In certain situations, the soil-dwelling amoeba *Dictyostelium* manages an extraordinary feat of social organization. When food is scarce individual cells – usually antisocial and free-living – clump together to form a

8. A MANIFESTO FOR LIFE

differentiated cell cluster known as a slug. Remarkably, a single molecule, *PKA*, is responsible for the whole process. Nematode worms, similarly, exhibit two naturally occurring foraging behaviours. Some individuals are solitary and feed alone; others are social and feed in aggregates. A single change in the gene *NPR-1* underlies this behavioural difference. There are many other examples of the genetic basis of social behaviour. Fire ants exhibit two types of social organization: there are monogyne colonies, which have only a single queen; or polygyne colonies, which can tolerate up to two hundred queens simultaneously. Their behaviour is determined by which of the two versions of a gene called *Gp-9* the colony possesses.

Genes influence many aspects of our behaviour, but there are instances where behaviour is situation-specific. It is not customary for humans to eat one another, but exceptional circumstances demand unusual responses. The Donner Party was a group of 81 migrants who became stranded in the Sierra Nevada in 1846. Their passage was halted by early snows and they were forced to set up camp for the winter. Their provisions ran out; eventually they were forced to eat their dead companions. There are other external factors that can control an organism's behaviour: microscopic parasites, for instance. Mosquitoes infected with the malaria-causing parasite *Plasmodium* behave oddly: what were happy-go-lucky airborne tourists become the insect equivalent of bloodthirsty Draculas. Infected mosquitoes aggressively seek out more encounters with other animals, increasing the parasite's chances of transmission, but risking death in action.

Despite the flexibility and plasticity of human culture, genes and their environmentally responsive programs have a more profound effect on our behaviour and mental world than we might ever have imagined. If every aspect of our behaviour, thoughts and emotions is shaped to some extent by our genetic programming, artificial modifications of these programs should enable key aspects of ourselves – including our shape, lifespan, intelligence, sense of equality, capacity for compassion, love, sexuality, empathy, aesthetics, justice and morality, all once assumed to be inviolable aspects of our humanity – to be modified or reconfigured from first principles. The mathematician Alan Turing – who first conceived of the idea of an 'electronic brain', later manifested as the digital computer – once addressed the issue of whether machines would ever be able to think and be conscious. Having posed the question he soon dismissed it, arguing that if a machine could behave 'as if' it were conscious, and could fool observers, the issue of whether or not it actually was conscious was nonsensical. After all, how else – except by observing its behaviour – are we to infer that another entity is conscious?

Should we attempt to remodel ourselves? We could argue – in the short term at least – that it is precisely the paradoxical and irrational aspects of our existence which make us so uniquely and charmingly human. The

8. A MANIFESTO FOR LIFE

philosopher David Hume argued that our negative attributes (our propensity for greed, warfare, excessive ambition, dishonesty, indecision, lust, anger), coupled with life's inescapable blights (illness, suffering, frailty, mortality), form an essential part of what defines us. From this perspective, modifications might fundamentally change our nature and might as a result be undesirable. We would also be running the risk of losing our capacity for free will and self-determination – the characteristic that perhaps more than anything else defines humanity.

Through greed, indifference, incompetence, incomplete knowledge, or intrinsic limitations in our ability to compute the consequences of our actions, we might undermine the continued existence of the natural world that sustains us. Species may be destroyed in mass extinctions; ecosystems may collapse. Communities that have been brought to a point of subtle balance by hundreds of thousands of years of shared evolutionary history may be destabilized by the accidental or deliberate introduction of synthetic life. We should remember that much of what we call nature has already been altered beyond all recognition. The highlands of Scotland, for example, are haunted by the ghosts of wolves, brown bears, reindeer, lynx, wild boars and beavers – all of which once lived there until they were destroyed by humans over the last few millennia.

But in the long term, the question of whether or not we should profoundly modify our nature, and that of other creatures, is, like Turing's question, absurd. It is inevitable that someone somewhere will eventually create advanced synthetic life and modify human nature beyond all recognition. The intervening ethical and philosophical issues are important details – perhaps the most important that mankind will ever have to consider – but they are details nonetheless. Synthetic life is inevitable because we are intrinsically curious, because we have utopian desires: these are inalienably human characteristics. We will not be able to pass up opportunities that allow us to feed the ever-increasing world population as crop yields reach their theoretical limits, to eliminate illness, to make onions that do not make us cry, to live longer, to give children to sterile parents – or to counter ideologies that threaten our way of life. Inevitability, however, is not an excuse for complacency.

As a fraction of the span of life's history and likely future duration, the timescale of ethical debates will lie within the realm of noise. This is not to deny the importance of such debates, but simply to acknowledge, for better or worse, the inevitability of our synthetic biological future. The discussions may run for tens or even hundreds of years. But they will be nothing more than a prelude, a finite and transient stepping-stone into a different type of future.

Homo sapiens has existed for around 130,000 years and was preceded by several very different species of humans – *Homo rudolphensis*, *Homo habilis*, *Homo erectus*, *Homo ergaster* – each coming and going at a rate of one every 200,000 years or so. Each no doubt had its own perspective on human existence. It would have been hard

8. A MANIFESTO FOR LIFE

for any of these human prototypes to have imagined that human existence could be configured in any other way, even if their mental capacities had been sufficiently complex to do so. It is very likely that there are kinds of human cognition and conscious experience unimaginable to us, ways of being that would make our own experience seem paltry and unsatisfactory. We cannot comprehend these things, just as a cat cannot imagine what it would be like to be human. The largely irrational urge to preserve our current incarnation unchanged is no different from wanting to keep red telephone boxes or milkmen. Human nature is not static: there is no rational basis for glorifying one aspect of life's evolutionary history and declaring it sacrosanct. The familiar is comforting – but then our instinct to avoid ourselves changing is no doubt itself to some extent genetically programmed.

The unlimited heredity systems of life almost certainly began using RNA technology, before undergoing a technological revolution, upgrading to DNA. It is possible that the age of DNA technology will in turn eventually be transcended, or at least profoundly modified. The information of life, this time encoded by artificial means, will explore combinatorial landscapes more powerful than those of RNA and DNA. These will be littered with unimaginable possibilities. We may need to accept that humans and all other DNA-based life are not an end point in themselves, but contingent beginnings: crucibles for the burning, baking and creation of life's future incarnations.

The author of *Mixing in Society*, who in 1874 dared readers to 'venture to predict the programme of an evening party in the year of grace 1969', wondered whether gentlemen would 'wear peaked shoes of such fantastic length that the points thereof shall be fastened to their girdles'. He also speculated whether 'our ladies, who have so recently warn an artificial cushion attached to the backs of their dresses' would by that time have 'transferred it to the tops of their heads'. If evening parties still exist in 2169, we should not be too surprised to find ourselves sitting next to a mermaid or a centaur, at a dinner hosted by a 159-year-old human with thin diaphanous wings and insect eyes. We might sit there and laugh at the irrational behaviours of our predecessors – who fell in love, lost their tempers, experienced jealousy, collected stamps, got ill, lived for only eighty years, and on occasion were discontented or unhappy. We might alternatively be incapable of imagining what such things were.

Bibliography

Abbott, A. With your genes? Take one of these, three times a day. *Nature* **425**, 760-762 (2003).

Agathon, A., Thisse, C. & Thisse, B. The molecular nature of the zebrafish tail organizer. *Nature* **424**, 448-452 (2003).

Almaas, E., Kovacs, B., Vicsek, T., Oltvai, Z. N. & Barabasi, A. L. Global organization of metabolic fluxes in the bacterium Escherichia coli. *Nature* **427**, 839-843 (2004).

Andersson, J. O., Doolittle, W. F. & Nesbo, C. L. Genomics. Are there bugs in our genome? *Science* **292**, 1848-1850 (2001).

Aparicio, S. et al. Whole-genome shotgun assembly and analysis of the genome of Fugu rubripes. *Science* **297**, 1301-1310 (2002).

Arbeitman, M. N., Furlong, E. E., Imam, F., Johnson, E., Null, B. H., Baker, B. S., Krasnow, M. A., Scott, M. P., Davis, R. W. & White, K. P. Gene expression during the life cycle of Drosophila melanogaster. *Science* **297**, 2270-2275 (2002).

Arthur, W. The emerging conceptual framework of evolutionary developmental biology. *Nature* **415**, 757-764 (2002).

Atkins, J. F. & Gesteland, R. Biochemistry. The 22nd amino acid. *Science* **296**, 1409-1410 (2002).

Audic, S. & Beraud-Colomb, E. Ancient DNA is thirteen years old. *Nat Biotechnol* **15**, 855-858 (1997).

Bairoch, A. & Murzin, A. G. Sequences and topology: predicting evolution. *Curr Opin Struct Biol* **7**, 367-368 (1997).

Balter, M. In Toulouse, the weather – and the science – are hot. *Science* **269**, 480-481 (1995).

Balter, M. Genetics. First gene linked to speech identified. *Science* **294**, 32 (2001).

Balter, M. Language, brain, and cognitive development meeting. What makes the mind dance and count. *Science* **292**, 1636-1637 (2001).

Balter, M. Paleoanthropology. What – or who – did in the Neanderthals? *Science* **293**, 1980-1981 (2001).

Balter, M. Language evolution. 'Speech gene' tied to modern humans. *Science* **297**, 1105 (2002).

Barlow, D. P. Gametic imprinting in mammals. *Science* **270**, 1610-1613 (1995).

Baulcombe, D. DNA events. An RNA microcosm. *Science* **297**, 2002-2003 (2002).

Beardsley, T. Smart genes. *Sci Am* **265**, 86-95 (1991).

Beckman, M. Pheromone reception. When in doubt, mice mate rather than hate. *Science* **295**, 782 (2002).

Beldade, P., Koops, K. & Brakefield, P. M. Developmental constraints versus flexibility in morphological evolution. *Nature* **416**, 844-847 (2002).

Benfey, P. N. Molecular biology: microRNA is here to stay. *Nature* **425**, 244-245 (2003).

Bergman, A. & Siegal, M. L. Evolutionary capacitance as a general feature of complex gene networks. *Nature* **424**, 549-552 (2003).

Bestor, T. H. Gene silencing. Methylation meets acetylation. *Nature* **393**, 311-312 (1998).

Blackwood, E. M. & Kadonaga, J. T. Going the distance: a current view of enhancer action. *Science* **281**, 60-63 (1998).

Blanke, O., Ortigue, S., Landis, T. & Seeck, M. Stimulating illusory own-body perceptions. *Nature* **419**, 269-270 (2002).

Blaxter, M. Comparative genomics: two worms are better than one. *Nature* **426**, 395-396 (2003).

Bloom, B. R. Genome sequences. A microbial minimalist. *Nature* **378**, 236 (1995).

Bloom, P. *Descartes' baby. How the science of child development explains what makes us human* (William Heinemann, London, 2004).

Bock, A. Molecular biology. Invading the genetic code. *Science* **292**, 453-454 (2001).

Boesch, C. The question of culture. *Nature* **379**, 207-208 (1996).

Boesch, C. Breaking down the barriers. *Nature* **411**, 525-526 (2001).

Boguski, M. S. Comparative genomics: the mouse that roared. *Nature* **420**, 515-516 (2002).

Bonhoeffer, S. & Sniegowski, P. Virus evolution: the importance of being erroneous. *Nature* **420**, 367, 369 (2002).

Borges, J. L. *Labyrinths* (eds Yates, D. A. & Irby, J. E.) (Penguin Books, London, 1970).

Bown, N. *Fairies in nineteenth-century art and literature* (Cambridge University Press, Cambridge, 2001).

BIBLIOGRAPHY

Bradley, A. Mining the mouse genome. *Nature* **420**, 512-514 (2002).

Bradley, D. Informatics. The genome chose its alphabet with care. *Science* **297**, 1789-1791 (2002).

Bray, D. Genomics. Molecular prodigality. *Science* **299**, 1189-1190 (2003).

Brenner, S. Loose ends. Centaur biology. *Curr Biol* **7**, R454 (1997).

Brenner, S. Interview with Sydney Brenner. The world of genome projects. *Bioessays* **18**, 1039-1042 (1996).

Brenner, S. Biological computation. *Novartis Found Symp* **213**, 106-111; discussion 111-106 (1998).

Brenner, S. Humanity as the model system. *Science* **302**, 533 (2003).

Brown, J. R., Ye, H., Bronson, R. T., Dikkes, P. & Greenberg, M. E. A defect in nurturing in mice lacking the immediate early gene fosB. *Cell* **86**, 297-309 (1996).

Bushman, F. Gene regulation: selfish elements make a mark. *Nature* **429**, 253-255 (2004).

Caplan, A. Bioethics. Is biomedical research too dangerous to pursue? *Science* **303**, 1142 (2004).

Carlson, E. A. *Mendel's legacy: the origin of classical genetics* (Cold Spring Harbor Laboratory Press, 2004).

Carroll, S. B. Homeotic genes and the evolution of arthropods and chordates. *Nature* **376**, 479-485 (1995).

Carroll, S. B. Genetics and the making of Homo sapiens. *Nature* **422**, 849-857 (2003).

Caspi, A., Sugden, K., Moffitt, T. E., Taylor, A., Craig, I. W., Harrington, H., McClay, J., Mill, J., Martin, J., Braithwaite, A. & Poulton, R. Influence of life stress on depression: moderation by a polymorphism in the 5-HTT gene. *Science* **301**, 386-389 (2003).

Cech, T. R. RNA finds a simpler way. *Nature* **428**, 263-264 (2004).

Cedar, H. & Verdine, G. L. Gene expression. The amazing demethylase. *Nature* **397**, 568-569 (1999).

Chargaff, E. How genetics got a chemical education. *Ann N Y Acad Sci* **325**, 344-360 (1979).

Check, E. Environmental impact tops list of fears about transgenic animals. *Nature* **418**, 805 (2002).

Check, E. Venter aims for maximum impact with minimal genome. *Nature* **420**, 350 (2002).

Check, E. Poliovirus advance sparks fears of data curbs. *Nature* **418**, 265 (2002).

Chen, R. Z., Pettersson, U., Beard, C., Jackson-Grusby, L. & Jaenisch,

R. DNA hypomethylation leads to elevated mutation rates. *Nature* **395**, 89-93 (1998).

Chenn, A. Making a bigger brain by regulating cell cycle exit. *Science* **298**, 766-767 (2002).

Chenn, A. Regulation of cerebral cortical size by control of cell cycle exit in neural precursors. *Science* **297**, 365-369 (2002).

Chin, J. W., Cropp, T. A., Anderson, J. C., Mukherji, M., Zhang, Z. & Schultz, P. G. An expanded eukaryotic genetic code. *Science* **301**, 964-967 (2003).

Cho, A. Life's patterns: no need to spell it out? *Science* **303**, 782-783 (2004).

Clark, A. G. et al. Inferring nonneutral evolution from human-chimp-mouse orthologous gene trios. *Science* **302**, 1960-1963 (2003).

Claverie, J. M. Gene number. What if there are only 30,000 human genes? *Science* **291**, 1255-1257 (2001).

Clayton, R. A., White, O., Ketchum, K. A. & Venter, J. C. The first genome from the third domain of life. *Nature* **387**, 459-462 (1997).

Cohen, J. Does nature drive nurture? *Science* **273**, 577-578 (1996).

Collins, F. S. & Watson, J. D. Genetic discrimination: time to act. *Science* **302**, 745 (2003).

Cossins, A. Cryptic clues revealed. *Nature* **396**, 309-310 (1998).

Couzin, J. Genomics. Painting a picture of genome evolution. *Science* **293**, 1969-1970 (2001).

Couzin, J. Genomics. Consensus emerges on HapMap strategy. *Science* **304**, 671-673 (2004).

Covert, M. W., Knight, E. M., Reed, J. L., Herrgard, M. J. & Palsson, B. O. Integrating high-throughput and computational data elucidates bacterial networks. *Nature* **429**, 92-96 (2004).

Crespi, B. & Springer, S. Ecology. Social slime molds meet their match. *Science* **299**, 56-57 (2003).

Currie, P. Human genetics: muscling in on hominid evolution. *Nature* **428**, 373-374 (2004).

Cyranoski, D. Almost human. *Nature* **418**, 910-912 (2002).

Czerny, T., Halder, G., Kloter, U., Souabni, A., Gehring, W. J. & Busslinger, M. Twin of eyeless, a second Pax-6 gene of Drosophila, acts upstream of eyeless in the control of eye development. *Mol Cell* **3**, 297-307 (1999).

Dalton, R. California edges towards farming drug-producing rice. *Nature* **428**, 591 (2004).

Davis, B. G. Biochemistry. Mimicking posttranslational modifications of proteins. *Science* **303**, 480-482 (2004).

Dawkins, R. *The blind watchmaker* (Penguin Books, London, 1988).

de Bono, M., Tobin, D. M., Davis, M. W., Avery, L. & Bargmann, C. I. Social feeding in Caenorhabditis elegans is induced by neurons that detect aversive stimuli. *Nature* **419**, 899-903 (2002).

De Robertis, E. M., Oliver, G. & Wright, C. V. Homeobox genes and the vertebrate body plan. *Sci Am* **263**, 46-52 (1990).

Dehaene, S. Neuroscience. Single-neuron arithmetic. *Science* **297**, 1652-1653 (2002).

Delbrück, M. A physicist's renewed look at biology – twenty years later. *Nobel Lecture* (December 10, 1969).

Dennett, D. C. in *The behavioural and brain sciences* 343-390 (Cambridge University Press, Cambridge, 1983).

Dennett, D. C. *Darwin's dangerous idea. Evolution and the meanings of life* (Simon & Schuster, New York, 1995).

Dennis, C. Mouse genome: a forage in the junkyard. *Nature* **420**, 458-459 (2002).

Dennis, C. Genomics: compare and contrast. *Nature* **426**, 750-751 (2003).

Dennis, C. The rough guide to the genome. *Nature* **425**, 758-759 (2003).

Dennis, C. Coral reveals ancient origins of human genes. *Nature* **426**, 744 (2003).

DeRisi, J. L., Iyer, V. R. & Brown, P. O. Exploring the metabolic and genetic control of gene expression on a genomic scale. *Science* **278**, 680-686 (1997).

Desmond, A. & Moore, J. *Darwin* (Penguin Books, London, 1991).

Diamond, J. *The rise and fall of the third chimpanzee* (Radius, 1991).

Diamond, J. Location, location, location: the first farmers. *Science* **278**, 1243-1243 (1997).

Dillon, N. Gene autonomy: positions, please. *Nature* **425**, 457 (2003).

Dolan, R. J. Emotion, cognition, and behavior. *Science* **298**, 1191-1194 (2002).

Donley, J. M., Sepulveda, C. A., Konstantinidis, P., Gemballa, S. & Shadwick, R. E. Convergent evolution in mechanical design of lamnid sharks and tunas. *Nature* **429**, 61-65 (2004).

Doring, V., Mootz, H. D., Nangle, L. A., Hendrickson, T. L., de Crecy-Lagard, V., Schimmel, P. & Marliere, P. Enlarging the amino acid set of Escherichia coli by infiltration of the valine coding pathway. *Science* **292**, 501-504 (2001).

Dujon, B. The yeast genome project: what did we learn? *Trends Genet* **12**, 263-270 (1996).

Eigen, M. *Steps towards life. A perspective on evolution* (Oxford University Press, Oxford, 1992).

Emmons, S. W. Simple worms, complex genes. *Nature* **382**, 301-302 (1996).

Enard, W., Przeworski, M., Fisher, S. E., Lai, C. S., Wiebe, V., Kitano, T., Monaco, A. P. & Paabo, S. Molecular evolution of FOXP2, a gene involved in speech and language. *Nature* **418**, 869-872 (2002).

Enard, W., Khaitovich, P., Klose, J., Zollner, S., Heissig, F., Giavalisco, P., Nieselt-Struwe, K., Muchmore, E., Varki, A., Ravid, R., Doxiadis, G. M., Bontrop, R. E. & Paabo, S. Intra- and interspecific variation in primate gene expression patterns. *Science* **296**, 340-343 (2002).

Erwin, D. H. The Goldilocks hypothesis. *Science* **302**, 1682-1683 (2003).

Fares, M. A., Ruiz-Gonzalez, M. X., Moya, A., Elena, S. F. & Barrio, E. Endosymbiotic bacteria: groEL buffers against deleterious mutations. *Nature* **417**, 398 (2002).

Fedoroff, N. Transposons and genome evolution in plants. *Proc Natl Acad Sci USA* **97**, 7002-7007 (2000).

Fedoroff, N. How jumping genes were discovered. *Nat Struct Biol* **8**, 300-301 (2001).

Fedoroff, N. V. Agriculture. Prehistoric GM corn. *Science* **302**, 1158-1159 (2003).

Fehr, E. & Fischbacher, U. The nature of human altruism. *Nature* **425**, 785-791 (2003).

Ferber, D. Synthetic biology. Microbes made to order. *Science* **303**, 158-161 (2004).

Ferster, D. Neuroscience. Blocking plasticity in the visual cortex. *Science* **303**, 1619-1621 (2004).

Fields, S. Proteomics. Proteomics in genomeland. *Science* **291**, 1221-1224 (2001).

Fitch, W. T. & Hauser, M. D. Computational constraints on syntactic processing in a nonhuman primate. *Science* **303**, 377-380 (2004).

Flam, F. Hints of a language in junk DNA. *Science* **266**, 1320 (1994).

Fortey, R. Evolution. The Cambrian explosion exploded? *Science* **293**, 438-439 (2001).

Fujiyama, A., Watanabe, H., Toyoda, A., Taylor, T. D., Itoh, T., Tsai, S. F., Park, H. S., Yaspo, M. L., Lehrach, H., Chen, Z., Fu, G., Saitou, N., Osoegawa, K., de Jong, P. J., Suto, Y., Hattori, M. &

Sakaki, Y. Construction and analysis of a human-chimpanzee comparative clone map. *Science* **295**, 131-134 (2002).

Galas, D. J. Sequence interpretation. Making sense of the sequence. *Science* **291**, 1257-1260 (2001).

Gaucher, E. A., Thomson, J. M., Burgan, M. F. & Benner, S. A. Inferring the palaeoenvironment of ancient bacteria on the basis of resurrected proteins. *Nature* **425**, 285-288 (2003).

Gee, H. *Jacob's ladder. The history of the human genome* (Fourth Estate, London, 2004).

Gellon, G. & McGinnis, W. Shaping animal body plans in development and evolution by modulation of Hox expression patterns. *Bioessays* **20**, 116-125 (1998).

Gerstein, M., Lan, N. & Jansen, R. Proteomics. Integrating interactomes. *Science* **295**, 284-287 (2002).

Giaever, G. et al. Functional profiling of the Saccharomyces cerevisiae genome. *Nature* **418**, 387-391 (2002).

Gibbons, A. Which of our genes makes us human? *Science* **281**, 1432-1434 (1998).

Gibbons, A. American Association of Physical Anthropologists meeting. Studying humans – and their cousins and parasites. *Science* **292**, 627-629 (2001).

Gibbons, A. Becoming human. In search of the first hominids. *Science* **295**, 1214-1219 (2002).

Gibbs, R. A. et al. The International HapMap Project. *Nature* **426**, 789-796 (2003).

Gibbs, R. A. et al. Genome sequence of the Brown Norway rat yields insights into mammalian evolution. *Nature* **428**, 493-521 (2004).

Goffeau, A. Life with 482 genes. *Science* **270**, 445-446 (1995).

Goldman, M. A. Promises and perils of technology's future. *Science* **303**, 629-630 (2004).

Goldsmith, T. H. Everyday impacts of a most influential theory. *Science* **293**, 2209-2210 (2001).

Gompel, N. & Carroll, S. B. Genetic mechanisms and constraints governing the evolution of correlated traits in drosophilid flies. *Nature* **424**, 931-935 (2003).

Gould, S. J. & Lewontin, R. C. The spandrels of San Marco and the Panglossian paradigm: a critique of the adaptationist programme. *Proc R Soc Lond B Biol Sci* **205**, 581-598 (1979).

Gould, S. J. The evolution of life on the earth. *Sci Am* **271**, 63-69 (1994).

Gould, S. J. Of it, not above it. *Nature* **377**, 681-682 (1995).

Gould, S. J. *The structure of evolutionary theory* (The Belknap Press of Harvard University Press, Harvard, 2002).

Gowaty, P. A. Behavioral just-so stories, recast. *Science* **293**, 610-611 (2001).

Gu, Z., Steinmetz, L. M., Gu, X., Scharfe, C., Davis, R. W. & Li, W. H. Role of duplicate genes in genetic robustness against null mutations. *Nature* **421**, 63-66 (2003).

Guet, C. C., Elowitz, M. B., Hsing, W. & Leibler, S. Combinatorial synthesis of genetic networks. *Science* **296**, 1466-1470 (2002).

Gunter, C. & Dhand, R. Human biology by proxy. *Nature* **420**, 509 (2002).

Gunturkun, O. Human behaviour: adult persistence of head-turning asymmetry. *Nature* **421**, 711 (2003).

Gura, T. One molecule orchestrates amoebae. *Science* **277**, 182 (1997).

Guss, K. A., Nelson, C. E., Hudson, A., Kraus, M. E. & Carroll, S. B. Control of a genetic regulatory network by a selector gene. *Science* **292**, 1164-1167 (2001).

Halder, G., Callaerts, P. & Gehring, W. J. Induction of ectopic eyes by targeted expression of the eyeless gene in Drosophila. *Science* **267**, 1788-1792 (1995).

Hamer, D. Genetics. Rethinking behavior genetics. *Science* **298**, 71-72 (2002).

Han, J. S., Szak, S. T. & Boeke, J. D. Transcriptional disruption by the L1 retrotransposon and implications for mammalian transcriptomes. *Nature* **429**, 268-274 (2004).

Hanada, K., Yewdell, J. W. & Yang, J. C. Immune recognition of a human renal cancer antigen through post-translational protein splicing. *Nature* **427**, 252-256 (2004).

Hariri, A. R., Mattay, V. S., Tessitore, A., Kolachana, B., Fera, F., Goldman, D., Egan, M. F. & Weinberger, D. R. Serotonin transporter genetic variation and the response of the human amygdala. *Science* **297**, 400-403 (2002).

Hartwell, L. Theoretical biology. A robust view of biochemical pathways. *Nature* **387**, 855, 857 (1997).

Hasty, J., McMillen, D. & Collins, J. J. Engineered gene circuits. *Nature* **420**, 224-230 (2002).

Hauser, M. D., Chomsky, N. & Fitch, W. T. The faculty of language: what is it, who has it, and how did it evolve? *Science* **298**, 1569-1579 (2002).

Hayes, W. Genetic transformation: a retrospective appreciation. *J Gen Microbiol* **45**, 385-397 (1966).

Hedges, S. B. & Kumar, S. Genomics. Vertebrate genomes compared. *Science* **297**, 1283-1285 (2002).

Helmuth, L. Society for Neuroscience meeting. Caudate-over-heels in love. *Science* **302**, 1320 (2003).

Henikoff, S., Greene, E. A., Pietrokovski, S., Bork, P., Attwood, T. K. & Hood, L. Gene families: the taxonomy of protein paralogs and chimeras. *Science* **278**, 609-614 (1997).

Henikoff, S. & Matzke, M. A. Exploring and explaining epigenetic effects. *Trends Genet* **13**, 293-295 (1997).

Hensch, T. K., Fagiolini, M., Mataga, N., Stryker, M. P., Baekkeskov, S. & Kash, S. F. Local GABA circuit control of experience-dependent plasticity in developing visual cortex. *Science* **282**, 1504-1508 (1998).

Higuchi, R., Bowman, B., Freiberger, M., Ryder, O. A. & Wilson, A. C. DNA sequences from the quagga, an extinct member of the horse family. *Nature* **312**, 282-284 (1984).

Hodgkin, J., Horvitz, H. R., Jasny, B. R. & Kimble, J. C. Elegans: sequence to biology. *Science* **282**, 2011 (1998).

Hoekstra, H. E. & Price, T. Evolution. Parallel evolution is in the genes. *Science* **303**, 1779-1781 (2004).

Hofreiter, M., Serre, D., Poinar, H. N., Kuch, M. & Paabo, S. Ancient DNA. *Nat Rev Genet* **2**, 353-359 (2001).

Holden, C. Animal behavior. Single gene dictates ant society. *Science* **294**, 1434 (2001).

Holden, C. Behavioral genetics. Getting the short end of the allele. *Science* **301**, 291-293 (2003).

Holden, C. Neuroscience. Deconstructing schizophrenia. *Science* **299**, 333-335 (2003).

Holder, N. & McMahon, A. Genes from zebrafish screens. *Nature* **384**, 515-516 (1996).

Holliday, R. A different kind of inheritance. *Sci Am* **260**, 60-65, 68-70, 73 (1989).

Holm, L. & Sander, C. Mapping the protein universe. *Science* **273**, 595-603 (1996).

Hunt, G. R., Corballis, M. C. & Gray, R. D. Animal behaviour: laterality in tool manufacture by crows. *Nature* **414**, 707 (2001).

Hutchison, C. A., Peterson, S. N., Gill, S. R., Cline, R. T., White, O., Fraser, C. M., Smith, H. O. & Venter, J. C. Global transposon mutagenesis and a minimal Mycoplasma genome. *Science* **286**, 2165-2169 (1999).

Huynen, M. A., Diaz-Lazcoz, Y. & Bork, P. Differential genome display. *Trends Genet* **13**, 389-390 (1997).

Ideker, T., Thorsson, V., Ranish, J. A., Christmas, R., Buhler, J., Eng, J. K., Bumgarner, R., Goodlett, D. R., Aebersold, R. & Hood, L. Integrated genomic and proteomic analyses of a systematically perturbed metabolic network. *Science* **292**, 929-934 (2001).

Imai, S., Tsuge, N., Tomotake, M., Nagatome, Y., Sawada, H., Nagata, T. & Kumagai, H. Plant biochemistry: an onion enzyme that makes the eyes water. *Nature* **419**, 685 (2002).

Jaenicke-Despres, V., Buckler, E. S., Smith, B. D., Gilbert, M. T., Cooper, A., Doebley, J. & Paabo, S. Early allelic selection in maize as revealed by ancient DNA. *Science* **302**, 1206-1208 (2003).

Jaenisch, R. DNA methylation and imprinting: why bother? *Trends Genet* **13**, 323-329 (1997).

Jasny, B. R. & Kennedy, D. The human genome. *Science* **291**, 1153 (2001).

Jayaraman, K. S. India debates results of its first transgenic cotton crop. *Nature* **421**, 681 (2003).

Johnson, N. A. Also Sprach Homo mutants. *Science* **294**, 1659 (2001).

Johnson, S. *Emergence. The connected lives of ants, brains, cities and software* (Allen Lane The Penguin Press, London, 2001).

Johnston, M. & Stormo, G. D. Evolution. Heirlooms in the attic. *Science* **302**, 997-999 (2003).

Jones, P. A. & Gonzalgo, M. L. Altered DNA methylation and genome instability: a new pathway to cancer? *Proc Natl Acad Sci USA* **94**, 2103-2105 (1997).

Jones, D. T. Structural biology. Learning to speak the language of proteins. *Science* **302**, 1347-1348 (2003).

Kahn, P. & Gibbons, A. DNA from an extinct human. *Science* **277**, 176-178 (1997).

Kaiser, J. When is a coho salmon not a coho salmon? *Science* **294**, 1806-1807 (2001).

Kauffmann, S. *At home in the universe. The search for laws of complexity* (Viking, London, 1995).

Kauffmann, S. Evolving evolvability. *Nature* **382**, 309-311 (1996).

Kaufman, T. C., Severson, D. W. & Robinson, G. E. The Anopheles genome and comparative insect genomics. *Science* **298**, 97-98 (2002).

Kawahara, Y., Ito, K., Sun, H., Aizawa, H., Kanazawa, I. & Kwak, S. Glutamate receptors: RNA editing and death of motor neurons. *Nature* **427**, 801 (2004).

BIBLIOGRAPHY

Keeling, P. J. Parasites go the full monty. *Nature* **414**, 401-402 (2001).

Kellis, M., Patterson, N., Endrizzi, M., Birren, B. & Lander, E. S. Sequencing and comparison of yeast species to identify genes and regulatory elements. *Nature* **423**, 241-254 (2003).

Kellis, M., Birren, B. W. & Lander, E. S. Proof and evolutionary analysis of ancient genome duplication in the yeast Saccharomyces cerevisiae. *Nature* **428**, 617-624 (2004).

Kennedy, D. The importance of rice. *Science* **296**, 13 (2002).

Kerr, R. A. Did Darwin get it all right? *Science* **267**, 1421-1422 (1995).

Kidwell, M. G. Keeping pace with opportunistic DNA. *Science* **295**, 2219-2220 (2002).

Kitano, H. Computational systems biology. *Nature* **420**, 206-210 (2002).

Klein, R. G. Paleoanthropology. Whither the Neanderthals? *Science* **299**, 1525-1527 (2003).

Knight, J. Biology's last taboo. *Nature* **413**, 12-15 (2001).

Knight, J. When the chips are down. *Nature* **410**, 860-861 (2001).

Koch, C. & Laurent, G. Complexity and the nervous system. *Science* **284**, 96-98 (1999).

Koonin, E. V., Wolf, Y. I. & Karev, G. P. The structure of the protein universe and genome evolution. *Nature* **420**, 218-223 (2002).

Koshland, D. E., Jr. Molecule of the year: the DNA repair enzyme. *Science* **266**, 1925 (1994).

Krieger, M. J. & Ross, K. G. Identification of a major gene regulating complex social behavior. *Science* **295**, 328-332 (2002).

Kuhlman, B., Dantas, G., Ireton, G. C., Varani, G., Stoddard, B. L. & Baker, D. Design of a novel globular protein fold with atomic-level accuracy. *Science* **302**, 1364-1368 (2003).

Lai, C. S. L., Fisher, S. E., Hurst, J. A., Vargha-Khadem, F. & Monaco, A. P. A forkhead-domain gene is mutated in a severe speech and language disorder. *Nature* **413**, 519-523 (2001).

Leader-Williams, N., Smith, R. J. & Walpole, M. J. Elephant hunting and conservation. *Science* **293**, 2203-2204 (2001).

Lee, T. I., Rinaldi, N. J., Robert, F., Odom, D. T., Bar-Joseph, Z., Gerber, G. K., Hannett, N. M., Harbison, C. T., Thompson, C. M., Simon, I., Zeitlinger, J., Jennings, E. G., Murray, H. L., Gordon, D. B., Ren, B., Wyrick, J. J., Tagne, J. B., Volkert, T. L., Fraenkel, E., Gifford, D. K. & Young, R. A. Transcriptional regulatory networks in Saccharomyces cerevisiae. *Science* **298**, 799-804 (2002).

Leshner, A. Science and sustainability. *Science* **297**, 897 (2002).

Levine, M. How insects lose their limbs. *Nature* **415**, 848-849 (2002).

Levinton, J. S. The big bang of animal evolution. *Sci Am* **267**, 84-91 (1992).

Lewin, B. The mystique of epigenetics. *Cell* **93**, 301-303 (1998).

Lim, L. P., Glasner, M. E., Yekta, S., Burge, C. B. & Bartel, D. P. Vertebrate microRNA genes. *Science* **299**, 1540 (2003).

Lindahl, T. Facts and artifacts of ancient DNA. *Cell* **90**, 1-3 (1997).

Lindblad-Toh, K. Genome sequencing: three's company. *Nature* **428**, 475-476 (2004).

Liu, H., Gao, J., Lynch, S. R., Saito, Y. D., Maynard, L. & Kool, E. T. A four-base paired genetic helix with expanded size. *Science* **302**, 868-871 (2003).

Lynch, M. & Conery, J. S. The origins of genome complexity. *Science* **302**, 1401-1404 (2003).

McClintock, B. Barbara McClintock's speech at the Nobel Banquet (December 10, 1983).

McClintock, B. The significance of responses of the genome to challenge. Nobel Lecture (December 8, 1983).

McComb, K., Moss, C., Durant, S. M., Baker, L. & Sayialel, S. Matriarchs as repositories of social knowledge in African elephants. *Science* **292**, 491-494 (2001).

McGinnis, W. & Kuziora, M. The molecular architects of body design. *Sci Am* **270**, 58-61, 64-56 (1994).

McMurray, M. A. & Gottschling, D. E. An age-induced switch to a hyper-recombinational state. *Science* **301**, 1908-1911 (2003).

Mann, C. C. Anthropology. Cracking the khipu code. *Science* **300**, 1650-1651 (2003).

Markl, H. Research doesn't denigrate humanity. *Nature* **412**, 479-480 (2001).

Marshall, E. The human genome. Sharing the glory, not the credit. *Science* **291**, 1189-1193 (2001).

Marshall, E. Genetics. Venter gets down to life's basics. *Science* **298**, 1701 (2002).

Marshall, E. Plant genetics. A hidden Arabidopsis emerges under stress. *Science* **296**, 1218 (2002).

Marshall Graves, J. A. The tree of life: view from a twig. *Science* **300**, 1621 (2003).

Masip, L., Pan, J. L., Haldar, S., Penner-Hahn, J. E., DeLisa, M. P., Georgiou, G., Bardwell, J. C. & Collet, J. F. An engineered pathway for the formation of protein disulfide bonds. *Science* **303**, 1185-1189 (2004).

BIBLIOGRAPHY

Mauroy, B., Filoche, M., Weibel, E. R. & Sapoval, B. An optimal bronchial tree may be dangerous. *Nature* **427**, 633-636 (2004).

Mayhew, H. *London labour and the London poor* (Penguin Books, London, 1985).

Meyer, A. Hox gene variation and evolution. *Nature* **391**, 225-228 (1998).

Meyer, A. Whither evo-devo? Assessing the interplay between evolutionary and developmental processes. *Nature* **413**, 455-457 (2001).

Meyer, A. Molecular evolution: duplication, duplication. *Nature* **421**, 31-32 (2003).

Miklos, G. L. & Rubin, G. M. The role of the genome project in determining gene function: insights from model organisms. *Cell* **86**, 521-529 (1996).

Milo, R., Shen-Orr, S., Itzkovitz, S., Kashtan, N., Chklovskii, D. & Alon, U. Network motifs: simple building blocks of complex networks. *Science* **298**, 824-827 (2002).

Mitchell-Olds, T. & Knight, C. A. Evolution. Chaperones as buffering agents? *Science* **296**, 2348-2349 (2002).

Mlot, C. Microbes hint at a mechanism behind punctuated evolution. *Science* **272**, 1741 (1996).

Moffat, A. S. Plant genetics. Finding new ways to protect drought-stricken plants. *Science* **296**, 1226-1229 (2002).

Monaco, A. P. A recipe for the mind. *Nature* **427**, 681 (2004).

Moore, M. J. Gene expression. When the junk isn't junk. *Nature* **379**, 402-403 (1996).

Morell, V. How the malaria parasite manipulates its hosts. *Science* **278**, 223 (1997).

Morris, S. C. *Life's solution. Inevitable humans in a lonely universe* (Cambridge University Press, Cambridge, 2003).

Moxon, E. R. & Thaler, D. S. Microbial genetics. The tinkerer's evolving tool-box. *Nature* **387**, 659, 661-652 (1997).

Mundy, N. I., Badcock, N. S., Hart, T., Scribner, K., Janssen, K. & Nadeau, N. J. Conserved genetic basis of a quantitative plumage trait involved in mate choice. *Science* **303**, 1870-1873 (2004).

Munte, T. F. Brains out of tune. *Nature* **415**, 589-590 (2002).

Murphy, M. P. & O'Neill, L. A. J. (eds) *What is life? The next fifty years* (Cambridge University Press, Cambridge, 1995).

Mushegian, A. R. & Koonin, E. V. A minimal gene set for cellular life derived by comparison of complete bacterial genomes. *Proc Natl Acad Sci USA* **93**, 10268-10273 (1996).

Myers, N. A convincing call for conversation. *Science* **295**, 447-448 (2002).

Nelson, L. S., Rosoff, M. L. & Li, C. Disruption of a neuropeptide gene, flp-1, causes multiple behavioral defects in Caenorhabditis elegans. *Science* **281**, 1686-1690 (1998).

Niehrs, C. Developmental biology: a tale of tails. *Nature* **424**, 375-376 (2003).

Nobrega, M. A., Ovcharenko, I., Afzal, V. & Rubin, E. M. Scanning human gene deserts for long-range enhancers. *Science* **302**, 413 (2003).

Normile, D. Building working cells 'in silico'. *Science* **284**, 80-81 (1999).

Normile, D. & Pennisi, E. The rice genome. Rice: boiled down to bare essentials. *Science* **296**, 32-36 (2002).

Nowak, R. Mining treasures from 'junk DNA'. *Science* **263**, 608-610 (1994).

Nurse, P. Systems biology: understanding cells. *Nature* **424**, 883 (2003).

Osborne, K. A., Robichon, A., Burgess, E., Butland, S., Shaw, R. A., Coulthard, A., Pereira, H. S., Greenspan, R. J. & Sokolowski, M. B. Natural behavior polymorphism due to a cGMP-dependent protein kinase of Drosophila. *Science* **277**, 834-836 (1997).

Paabo, S. Genomics and society. The human genome and our view of ourselves. *Science* **291**, 1219-1220 (2001).

Palatnik, J. F., Allen, E., Wu, X., Schommer, C., Schwab, R., Carrington, J. C. & Weigel, D. Control of leaf morphogenesis by microRNAs. *Nature* **425**, 257-263 (2003).

Patthy, L. *Protein evolution* (Blackwell Science, Oxford, 1999).

Pennisi, E. Tracing backbone evolution through a tunicate's lost tail. *Science* **274**, 1082-1083 (1996).

Pennisi, E. Chemical shackles for genes? *Science* **273**, 574-575 (1996).

Pennisi, E. Seeking life's bare (genetic) necessities. *Science* **272**, 1098-1099 (1996).

Pennisi, E. & Roush, W. Developing a new view of evolution. *Science* **277**, 34-37 (1997).

Pennisi, E. Polymer folds just like a protein. *Science* **277**, 1764 (1997).

Pennisi, E. Heat shock protein mutes genetic changes. *Science* **282**, 1796 (1998).

Pennisi, E. Behind the scenes of gene expression. *Science* **293**, 1064-1067 (2001).

BIBLIOGRAPHY

Pennisi, E. Molecular evolution. Genome duplications: the stuff of evolution? *Science* **294**, 2458-2460 (2001).

Pennisi, E. Evolutionary genetics. Jumbled DNA separates chimps and humans. *Science* **298**, 719-721 (2002).

Pennisi, E. Evolutionary biology. Evo-devo enthusiasts get down to details. *Science* **298**, 953-955 (2002).

Pennisi, E. Primate evolution. Gene activity clocks brain's fast evolution. *Science* **296**, 233-235 (2002).

Pennisi, E. Genetics. One gene determines bee social status. *Science* **296**, 636 (2002).

Pennisi, E. Comparative genomics. Tunicate genome shows a little backbone. *Science* **298**, 2111-2112 (2002).

Pennisi, E. Comparative physiology. Recharged field's rallying cry: gene chips for all organisms. *Science* **297**, 1985-1987 (2002).

Pennisi, E. Evolution: Genome comparisons hold clues to human evolution. *Science* **302**, 1876-1877 (2003).

Pennisi, E. Molecular biology. Venter cooks up a synthetic genome in record time. *Science* **302**, 1307 (2003).

Pennisi, E. Bioinformatics. Gene counters struggle to get the right answer. *Science* **301**, 1040-1041 (2003).

Pennisi, E. Invertebrate conservation. Respect for things that flutter, creep, and crawl. *Science* **304**, 27 (2004).

Pennisi, E. Evolution of developmental diversity meeting. RNAi takes Evo-Devo world by storm. *Science* **304**, 384 (2004).

Penny, S. The pursuit of the living machine. *Sci Am,* 172 (1995).

Penny, D. Evolutionary biology: our relative genetics. *Nature* **427**, 208 (2004).

Pepper, H. The true history of the ghost and all about metempsychosis. (1890).

Piao, X., Hill, R. S., Bodell, A., Chang, B. S., Basel-Vanagaite, L., Straussberg, R., Dobyns, W. B., Qasrawi, B., Winter, R. M., Innes, A. M., Voit, T., Ross, M. E., Michaud, J. L., Déscarie, J.-C., Barkovich, A. J. & Walsh, C. A. G protein-coupled receptor-dependent development of human frontal cortex. *Science* **303**, 2033-2036 (2004).

Pigliucci, M. Developmental genetics: buffer zone. *Nature* **417**, 598-599 (2002).

Pinker, S. Talk of genetics and vice versa. *Nature* **413**, 465-466 (2001).

Plotkin, J. B., Dushoff, J. & Fraser, H. B. Detecting selection using a single genome sequence of M. tuberculosis and P. falciparum. *Nature* **428**, 942-945 (2004).

Porter, R. *Flesh in the age of reason. The modern foundations of body and soul* (W. W. Norton & Company, New York, 2003).

Premack, D. Psychology. Is language the key to human intelligence? *Science* **303**, 318-320 (2004).

Quackenbush, J. Genomics. Microarrays – guilt by association. *Science* **302**, 240-241 (2003).

Queitsch, C., Sangster, T. A. & Lindquist, S. Hsp90 as a capacitor of phenotypic variation. *Nature* **417**, 618-624 (2002).

Raff, R. A. Shadows of Plato's genes. Regulatory gene pathways could explain animals' different morphologies. *Nature* **418**, 917-918 (2002).

Rakic, P. Neuroscience. Genetic control of cortical convolutions. *Science* **303**, 1983-1984 (2004).

Rammensee, H. G. Immunology: protein surgery. *Nature* **427**, 203-204 (2004).

Rao, A. Sampling the universe of gene expression. *Nat Biotechnol* **16**, 1311-1312 (1998).

Rasmussen, L. E., Lee, T. D., Roelofs, W. L., Zhang, A. & Daves, G. D., Jr. Insect pheromone in elephants. *Nature* **379**, 684 (1996).

Rasmussen, S., Chen, L., Deamer, D., Krakauer, D. C., Packard, N. H., Stadler, P. F. & Bedau, M. A. Evolution. Transitions from non-living to living matter. *Science* **303**, 963-965 (2004).

Ravasz, E., Somera, A. L., Mongru, D. A., Oltvai, Z. N. & Barabasi, A. L. Hierarchical organization of modularity in metabolic networks. *Science* **297**, 1551-1555 (2002).

Raven, P. H. Presidential address. Science, sustainability, and the human prospect. *Science* **297**, 954-958 (2002).

Reeves, R. H. Functional genomics: a time and place for every gene. *Nature* **420**, 518-519 (2002).

Reik, W. & Dean, W. Back to the beginning. *Nature* **420**, 127 (2002).

Richardson, M. K. & Brakefield, P. M. Developmental biology: hotspots for evolution. *Nature* **424**, 894-895 (2003).

Riddihough, G. & Pennisi, E. The evolution of epigenetics. *Science* **293**, 1063 (2001).

Roberts, D. L. & Solow, A. R. Flightless birds: when did the dodo become extinct? *Nature* **426**, 245 (2003).

Roberts, L. The human genome. Controversial from the start. *Science* **291**, 1182-1188 (2001).

Robinson, G. E. Development. Sociogenomics takes flight. *Science* **297**, 204-205 (2002).

Robinson, G. E. Genomics. Beyond nature and nurture. *Science* **304**, 397-399 (2004).

Roemer, I., Reik, W., Dean, W. & Klose, J. Epigenetic inheritance in the mouse. *Curr Biol* **7**, 277-280 (1997).

Roush, W. A new embryo zoo. *Science* **274**, 1608-1609 (1996).

Roush, W. 'Smart' genes use many cues to set cell fate. *Science* **272**, 652-653 (1996).

Roush, W. Developmental biology. Corn: a lot of change from a little DNA. *Science* **272**, 1873 (1996).

Roush, W. Evolutionary biology: sizing up dung beetle evolution. *Science* **277**, 184 (1997).

Rutherford, S. L. & Lindquist, S. Hsp90 as a capacitor for morphological evolution. *Nature* **396**, 336-342 (1998).

Sachs, J. D. Sustainable development. *Science* **304**, 649 (2004).

Salehi-Ashtiani, K. & Szostak, J. W. In vitro evolution suggests multiple origins for the hammerhead ribozyme. *Nature* **414**, 82-84 (2001).

Salzberg, S. L. Genomics: yeast rises again. *Nature* **423**, 233-234 (2003).

Sancar, A. DNA repair in humans. *Annu Rev Genet* **29**, 69-105 (1995).

Sanger, F. Nobel lecture: Determination of nucleotide sequences in DNA. *Chemistry,* 431-447 (8 December 1980).

Sanger, F. Frederick Sanger's speech at the Nobel Banquet (December 10, 1980).

Secord, J. A. Portraits of science. Quick and magical shaper of science. *Science* **297**, 1648-1649 (2002).

Sejnowski, T. J. A high point for evolution. *Science* **283**, 1121-1122 (1999).

Service, R. F. Rock paintings yield DNA. *Science* **268**, 501 (1995).

Service, R. F. Metabolic engineering. Researchers create first autonomous synthetic life form. *Science* **299**, 640 (2003).

Shapiro, B., Sibthorpe, D., Rambaut, A., Austin, J., Wragg, G. M., Bininda-Emonds, O. R., Lee, P. L. & Cooper, A. Flight of the dodo. *Science* **295**, 1683 (2002).

Shapiro, M. D., Marks, M. E., Peichel, C. L., Blackman, B. K., Nereng, K. S., Jonsson, B., Schluter, D. & Kingsley, D. M. Genetic and developmental basis of evolutionary pelvic reduction in threespine sticklebacks. *Nature* **428**, 717-723 (2004).

Shubin, N. H. & Dahn, R. D. Evolutionary biology: lost and found. *Nature* **428**, 703-704 (2004).

Slabbekoorn, H. & Peet, M. Ecology: birds sing at a higher pitch in urban noise. *Nature* **424**, 267 (2003).

Smith, J. M. Natural selection and the concept of a protein space. *Nature* **225**, 563-564 (1970).

Smith, J. M. & Szathmary, E. *The origins of life. From the birth of life to the origin of language* (Oxford University Press, Oxford, 1999).

Sniegowski, P. Evolution: setting the mutation rate. *Curr Biol* **7**, R487-488 (1997).

Sokolowski, M. B. Neurobiology: social eating for stress. *Nature* **419**, 893-894 (2002).

Spier, R. E. Toward a new human species? *Science* **296**, 1807-1809 (2002).

Stedman, H. H., Kozyak, B. W., Nelson, A., Thesier, D. M., Su, L. T., Low, D. W., Bridges, C. R., Shrager, J. B., Minugh-Purvis, N. & Mitchell, M. A. Myosin gene mutation correlates with anatomical changes in the human lineage. *Nature* **428**, 415-418 (2004).

Stephanopoulos, G. & Kelleher, J. Biochemistry. How to make a superior cell. *Science* **292**, 2024-2025 (2001).

Sterne, L. *The life and opinions of Tristram Shandy gentleman* (Wordsworth Editions Limited, Ware, Herts, 1996).

Stokstad, E. Psychology. Violent effects of abuse tied to gene. *Science* **297**, 752 (2002).

Stokstad, E. Animal biotechnology. Environmental impact seen as biggest risk. *Science* **297**, 1257 (2002).

Stokstad, E. Paleontology. Ancient DNA pulled from soil. *Science* **300**, 407 (2003).

Stone, R. DNA forensics. Buried, recovered, lost again? The Romanovs may never rest. *Science* **303**, 753 (2004).

Stowers, L., Holy, T. E., Meister, M., Dulac, C. & Koentges, G. Loss of sex discrimination and male-male aggression in mice deficient for TRP2. *Science* **295**, 1493-1500 (2002).

Stuart, J. M., Segal, E., Koller, D. & Kim, S. K. A gene-coexpression network for global discovery of conserved genetic modules. *Science* **302**, 249-255 (2003).

Sucena, E., Delon, I., Jones, I., Payre, F. & Stern, D. L. Regulatory evolution of shavenbaby/ovo underlies multiple cases of morphological parallelism. *Nature* **424**, 935-938 (2003).

Summers, A. P. Biomechanics: fast fish. *Nature* **429**, 31-33 (2004).

Sun, F. L., Dean, W. L., Kelsey, G., Allen, N. D. & Reik, W. Transactivation of Igf2 in a mouse model of Beckwith-Wiedemann syndrome. *Nature* **389**, 809-815 (1997).

Surridge, C. Plant development: leaves by number. *Nature* **426**, 237 (2003).

Szathmary, E. Evolution. Developmental circuits rewired. *Nature* **411**, 143-145 (2001).

Tatusov, R. L., Koonin, E. V. & Lipman, D. J. A genomic perspective on protein families. *Science* **278**, 631-637 (1997).

Thomas, J. W. et al. Comparative analyses of multi-species sequences from targeted genomic regions. *Nature* **424**, 788-793 (2003).

Thompson, D. *On growth and form* (Cambridge University Press, Cambridge, 1961).

Tilghman, S. M. The sins of the fathers and mothers: genomic imprinting in mammalian development. *Cell* **96**, 185-193 (1999).

Turing, A. M. Computing machinery and intelligence. *Mind* **59**, 433-460 (1950).

van Schaik, C. P., Ancrenaz, M., Borgen, G., Galdikas, B., Knott, C. D., Singleton, I., Suzuki, A., Utami, S. S. & Merrill, M. Orang-utan cultures and the evolution of material culture. *Science* **299**, 102-105 (2003).

Vandermeer, J. The importance of a constructivist view. *Science* **303**, 472-474 (2004).

Vauclair, J. Mental states in animals: cognitive ethology. *Trends Cog Sci* **1**, 35-39 (1997).

Vogel, G. DNA suggests cultural traits affect whales' evolution. *Science* **282**, 1616 (1998).

Vogel, G. Missized mutants help identify organ tailors. *Science* **297**, 328 (2002).

Vogel, G. Development. Mutations reveal genes in zebrafish. *Science* **296**, 1221 (2002).

Vogel, G. Animal behavior. Orang-utans, like chimps, heed the cultural call of the collective. *Science* **299**, 27-28 (2003).

von Neumann, J. *The computer and the brain* (Yale University Press, New Haven, 1958).

Wang, L., Brock, A., Herberich, B. & Schultz, P. G. Expanding the genetic code of Escherichia coli. *Science* **292**, 498-500 (2001).

Wang, L. Amersham Prize winner. Expanding the genetic code. *Science* **302**, 584-585 (2003).

Ward, R. & Stringer, C. A molecular handle on the Neanderthals. *Nature* **388**, 225-226 (1997).

Warren, L. Mrs Tom Thumb's autobiography. *New York Tribune Sunday Magazine* (September 16, 1906).

Watanabe, H. et al. DNA sequence and comparative analysis of chimpanzee chromosome 22. *Nature* **429**, 382-388 (2004).

Waterston, R. H. et al. Initial sequencing and comparative analysis of the mouse genome. *Nature* **420**, 520-562 (2002).

Watson, J. D. The involvement of RNA in the synthesis of proteins. Nobel Lecture (December 11, 1962).

Weir, A. A., Chappell, J. & Kacelnik, A. Shaping of hooks in New Caledonian crows. *Science* **297**, 981 (2002).

Weissenbach, J. Genome sequencing: differences with the relatives. *Nature* **429**, 353-355 (2004).

Whitfield, C. W., Cziko, A. M. & Robinson, G. E. Gene expression profiles in the brain predict behavior in individual honey bees. *Science* **302**, 296-299 (2003).

Whitfield, J. Origins of life: born in a watery commune. *Nature* **427**, 674-676 (2004).

Wilkins, M. H. F. The molecular configuration of nucleic acids. Nobel Lecture (December 11, 1962).

Winkler, W. C., Nahvi, A., Roth, A., Collins, J. A. & Breaker, R. R. Control of gene expression by a natural metabolite-responsive ribozyme. *Nature* **428**, 281-286 (2004).

Wood, B. Human evolution. *Bioessays* **18**, 945-954 (1996).

Wood, R. D., Mitchell, M., Sgouros, J. & Lindahl, T. Human DNA repair genes. *Science* **291**, 1284-1289 (2001).

Wood, B. So near, but yet so far. *Nature* **420**, 609 (2002).

Woodward, I. Plant science: tall storeys. *Nature* **428**, 807-808 (2004).

Woolfson, A. *Life without genes* (Flamingo, London, 2000).

Wray, G. A. Promoter logic. *Science* **279**, 1871-1872 (1998).

Wray, G. A. Resolving the Hox paradox. *Science* **292**, 2256-2257 (2001).

Yekta, S., Shih, I. H. & Bartel, D. P. MicroRNA-directed cleavage of HOXB8 mRNA. *Science* **304**, 594-596 (2004).

You, L., Cox, R. S., 3rd, Weiss, R. & Arnold, F. H. Programmed population control by cell-cell communication and regulated killing. *Nature* **428**, 868-871 (2004).

Yuh, C. H., Bolouri, H. & Davidson, E. H. Genomic cis-regulatory logic: experimental and computational analysis of a sea urchin gene. *Science* **279**, 1896-1902 (1998).

Zaslavskaia, L. A., Lippmeier, J. C., Shih, C., Ehrhardt, D., Grossman, A. R. & Apt, K. E. Trophic conversion of an obligate photoautotrophic organism through metabolic engineering. *Science* **292**, 2073-2075 (2001).

Zdobnov, E. M. et al. Comparative genome and proteome analysis of Anopheles gambiae and Drosophila melanogaster. *Science* **298**, 149-159 (2002).

Zeki, S. Essays on science and society. Artistic creativity and the brain. *Science* **293**, 51-52 (2001).

Zhang, Y. X., Perry, K., Vinci, V. A., Powell, K., Stemmer, W. P. & del Cardayre, S. B. Genome shuffling leads to rapid phenotypic improvement in bacteria. *Nature* **415**, 644-646 (2002).

Zimmer, C. Genomics. Tinker, tailor: can Venter stitch together a genome from scratch? *Science* **299**, 1006-1007 (2003).

Zuker, C. S. On the evolution of eyes: would you like it simple or compound? *Science* **265**, 742-743 (1994).

Index

acetylation, 100, 103
Adams, Mark, 83
Adelaide, Queen, 51
Alberts, Bruce, 81
Alexei, Tsarevich, 12
alternative materials, 155-6
alternative polyadenylation, 103
alternative splicing, 102
Ambros, Victor, 96
amino acids, 48-9, 140, 157-8, 186, 188, 192
  see also proteins
amoeba *Dictyostelium*, 202-3
amplification, DNA, 17-18
amyotrophic lateral sclerosis, 103
analytic biology, 129
ancestral gene sets, 23-4, 26-8, 131
ancient material and life forms, 12-16, 17, 19-28, 145
  see also dodos
annelids *see* worms
Apt, K.E., 139
Arabidopsis thaliana, 88, 89, 97, 121
Archytas of Tarentum, 127
Arctic skua, 173
Aristotle, 56, 57, 60
Arnold, Frances, 134
arthropods, 26, 122-3
artificial genomes *see* synthetic biology
attractor regions, 168, 174
automata, 125-8
Avalos y Figueroa, Diego, 177
Avery, J.D., 53
Avery, Oswald Theodore, 40-1, 42

Babbage, Charles, 126
Babington, Charles, 66
Bacon, Francis, 57-9
Bacon, Roger, 127

bacteria
  cell size study, 124
  DNA, 19, 24-5, 43-4, 84, 106, 162
  genes, 90, 130, 132-3, 183
  microRNAs (miRNAs), 97
  synthetic biology, 130, 132-3, 134, 135
  transformation experiments, 37-41
  transposons, 107, 110
  see also Escherichia coli; *Haemophilus influenzae*; hyperthermophilic bacteria
bacteriophages, 43-4, 79, 131-2
Baer, Ernst von, 63
Baker, David, 160
Barabasi, A.L., 144
Barnum, Phineas Taylor, 51, 52, 53
Barrell, Bart, 80
Bateson, William, 34, 67-9, 72, 108
Beckwith, John Bruce, 98
bees, 68-9, 142, 143
behaviour, 12, 202-6
  see also cultural inheritance; environmental factors
Bender, Welcome, 70
Bentinck, Count William, 62
Berlin, Isaiah, 147
Bermuda Rules, 87
Bernoulli principle, 150
bestiaries, 55-7, 66, 67-9, 72
beta-catenin protein, 193
bias, in evolution, 169, 174, 175-6
bilateral frontoparietal polymicrogyria (BFPP), 194
birds, 15, 27, 173-4
Black, John, 149
Blanke, Olaf, 184
Bloom, Paul, 184
bones, DNA extraction from, 12-16, 17, 19
Borges, Jorge Luis, 161

INDEX

Botstein, David, 80-1
Bowen, Eli, 53
Boyle, Robert, 146-7
Braddon, Mary, 148
brains, 101, 104, 128-9, 166, 180, 183-5, 192-201
   see also consciousness; language; mental phenomena; mind
Brenner, Sydney, 80, 81, 106
bronchial trees, 154-5
Brooks, Mary, 53
Brown, Jennifer, 202
Brown, Patrick, 142
bryozoa, 124
BWS (Beckwith-Wiedemann Syndrome), 98, 99

Callaerts, Patrick, 96
cancers, 98, 113
Capsi, Avshalom, 201
Cargill, Michele, 187
cave paintings, 20-1
Celera Genomics, 85, 86, 87, 88
cerebral cortex, 192-4
chance, role of, 168-9
Chargaff, Erwin, 37, 42, 45
Charles I, King, 60
Chase, Martha, 43, 44, 45
Cheetham, Alan, 124
chemical constraints, 150-1
Chen, Anjen, 193
Cheyne, George, 59
Childs, Avery, 53
chimpanzees, 167, 182-3, 186-8, 189, 192, 197, 198
Chomsky, Noam, 188
chromatins, 99-100
chromosomes, 106-7
   damage to, 108-10
   DNA and, 18, 99
   fruit flies, 35-6, 92
   genes and, 36, 108
   genetic information linked to, 34-7
   genomic churning, 115-16
   maize, 108-9
   microsatellite sequences and, 113
   recombination, 108
   retrotransposon effect on, 111
   telomeres, 115-16
   see also DNA; genes
circuits *see* genetic circuits
Clinton, President William J., 87
CNGs (conserved non-genic sequences), 115
codons, 49-50, 157
Coho salmon, 167
Cole, Rufus, 40
Collins, Francis, 84, 85, 86, 87, 88
Collins, Timothy, 124
complexity, 89-91, 93-4, 96, 98-102, 151, 161-2, 182
congenital amusia, 181-2
Conklin, William J., 178
connectivity, 138, 144, 173
consciousness, 138, 183-4, 185, 204
conserved non-genic sequences (CNGs), 115
constraints, 149-61, 169, 171-2, 174-6, 181, 190
constructional biology *see* synthetic biology
control, gene expression *see* regulation, gene expression
convergence, 15, 169-74
   see also parallel evolution
Copernicus, Nicolaus, 30
copying *see* duplication; replication
corals, 26
Correns, Carl, 34
cranial expansion, humans, 195-7
Crick, Francis, 45, 49, 69
Cromwell, Oliver, 60
cultural inheritance, 12, 141, 166-8, 177-82, 185-6
   see also language

Dante, 78
Darwin, Charles R., 29-31, 66-7, 175
Darwinism, 29-31, 32, 34, 65-7, 119, 121, 123-4
Davidson, Eric H., 92, 93
decay, DNA, 13, 16, 21
   see also stability, genetic systems
degeneracy, DNA sequences, 49
Delbrück, Max, 43
deletions, 26, 111, 115, 172, 186-7

INDEX

DeLisi, Charles, 81
Democritus, 57
depression, 201-2
Descartes, René, 78, 79, 183
developmental constraints, 152-3, 169, 171-2, 174-5
developmental drive, 174
developmental hotspots, 171-2, 173, 174
Devere, Madame, 53
Diamond, Jared, 182
Dickens, Charles, 76
diploid genes, 98
discontinuous evolution, 67-9
  see also saltatory transitions
disease susceptibility genes, 84, 188, 201
dissociation sites, 108-9
DNA, 9-28, 207
  bacterial, 43-4, 106, 162
  chromosomes and, 18, 99
  genes and, 18, 41-4, 46
  material constraint, 155
  microRNAs, 89
  protein synthesis and, 49
  replication, 16, 45
  retrotransposon effects on, 111
  RNA conversion to, 110
  structure, 11, 42, 45, 46
  synthetic, 156-7
  transformation phenomenon and, 40-1
  see also chromosomes; exons; genes; introns; mitochondria
DNA polymerase, 45, 46
  see also polymerase chain reaction technique
DNA sequences
  ancient/ancestral, 22-8, 131
  bacteria, 19, 24-5, 106, 162
  components of, 11
  computability, 135
  degeneracy, 49
  duplication, 115
  fixing mutations into, 24, 46, 50, 120, 171, 173
  gene families, 140-1
  gene inventories, 139-40
  genomic imprinting, 99, 100
  human vs. other species, 26, 82, 138, 182-3, 186-8, 192, 197-8
  inferences from, 23-5, 136, 162-3
  information storage, 11-12, 16, 46, 48-50, 107, 128-9, 162-9, 179
  insertions and deletions, 186
  mice, 19, 187-8
  minimal, 130-1, 161-7
  plants, 19, 96
  proteins and, 17, 18, 48-9
  retrotransposon effects on, 109-10, 111
  synthetic biology, 131-5, 138-40, 141, 143
  viruses, 19, 133, 162
  see also codons; DNA space; genes; homeobox; human genome; non-coding DNA sequences; nucleotides; regulatory sequences; transposons
DNA sequencing, 18-19, 79-88, 130, 139, 186
  human genome, 79-88, 99, 106, 107, 182, 186
DNA space, 163-6, 168, 169, 174, 175-6
Dodgson, Charles Lutwidge, 14
dodos, 14-15, 27, 161, 164, 166-7
Dollo's law, 159
double helix structure, 45
dung beetles, 152
Dunn, Leslie, 92
duplication, genes, 114-15

Edinburgh, Duke of, 13
Eldredge, Nils, 121
elephants, 167
Elizabeth I, Queen, 60
elongation factors, 25
embryology, 54, 60-2, 64-5, 72-4
Enard, Wolfgang, 197
Endo 16 gene promoter, 93-4, 136-7
enhancers, 93, 97, 101, 113, 141
environmental factors, 120, 165, 166-7, 174, 198-9, 201-2

INDEX

environmental inferences, 20, 22, 23, 24-5
Epicurus, 57
epigenesis, 61, 62-4
epigenetic information, 141, 165, 166
epigenetic modifications, 99, 110
epigenotypes, 99, 100
epimerization, 103
Epstein-Barr virus, 80
Escherichia coli, 44, 82, 134-5, 136, 144, 158
ESTs (expressed sequence tags), 83
Evertsz, Volkert, 15
evolution
    attractor regions, 168
    bias in, 169, 174, 175-6
    chance and, 168-9
    constraints on, 149-61, 169, 171-2, 174-6
    convergence, 15, 169-74
    environmental factors, 165, 174
    gene duplication, 114
    gene expression, 198
    genetic modules and, 138
    genetic reprogramming, 197-9
    genetic stability and variability, 46
    junk DNA and, 113-14
    master control genes' role, 118-19
    see also Darwinism; discontinuous evolution; gradualistic evolution; macroevolution; microevolution; parallel evolution; saltatory transitions
evolvability, 134, 144, 155
exons, 102
extinct life forms *see* ancient material and life forms
extragenetic information storage, 177-81
    see also brains
eyeless gene, flies, 95-6
eyesight, 153

Faber, Joseph, 125, 126-7
faecal samples, 20
fairies, 146-9, 161

families *see* genes, families of; proteins, families of
Faraday, Michael, 75
Fibonacci series, 150-1
fire ants, 203
fish, 19, 73, 88, 90, 140, 167, 169-72
Fitch, Tecumseh, 190
Fitzgerald, John Anster, 148
Fitzroy, Robert, 30
5-HTT gene, 201-2
Fleischmann, Rob, 84
flies
    bestiary of, 72
    chromosomes, 35-6, 92
    DNA sequences, 19, 82, 86
    gene expression, 94, 142
    gene families, 140
    gene methylation, 101
    genes, 26, 72-3, 89, 90, 92, 95-6, 171, 172
    homeotic variations, 69-70
    HSP90 (heat-shock protein), 120-1
    parallel evolution, 172
    'position effect' mutants, 92
    transcription factors, 95-6
folds, proteins, 120, 159-60
forkhead domain, proteins, 192
Fox, William Darwin, 66
Franklin, Rosalind, 45
Fraser, Claire, 84
Freind, J., 127
frogs, 70-1
frontal lobes, human brains, 193-4
fruit flies *see* flies
functional constraints, 154-5
Fuseli, Henry, 148

Galilei, Galileo, 30
gap genes, 72
gastrulation, 73-4
Gaucher, Eric, 24, 25
Gehring, Walter, 70, 95
GenBank, 87
gene, definition, 18, 89, 92
gene cassettes, bacteria, 132-3
gene duplication, 114-15, 159
    see also replication

INDEX

gene expression, 91-102, 109-10, 113, 136-7, 142-3, 173
  evolution, 198
  miRNAs, 89, 96-7, 101-2
  programs, 142-3, 197-8
  retrotransposons and, 109, 112
  sticklebacks, 170-2
  technologies, 142, 197-8
gene inventories, 139-40
gene loss *see* deletions
gene numbers, 86-91, 94, 130-1, 182
gene size, 90
gene technology, 141-2, 197-8
genes
  ancestral sets, 23-4, 131
  bacteria, 90, 130, 132-3, 183
  behaviour control, 200-3, 204
  chromosomes and, 36, 108
  culture and, 11-12, 181-2
  deacetylation of, 100
  discovery of, 31-8
  diseases and, 84, 188, 201
  DNA and, 18, 41-4, 46
  families of, 140-1
  flies, 26, 89, 90, 95-6, 171
  fruit flies, 72-3, 92, 172
  genetic modules and, 137
  homeotic complex, 70
  human, 88-90, 140, 141, 182, 183, 191-2
  inheritance of, 98-9
  language disorders, 191-2
  methylation of, 99-101, 166
  mice, 90, 97, 140, 201, 202
  non-coding RNA, 115
  proteins and, 91-2, 97, 100-1, 102
  retrotransposon effects on, 111
  smartness, 91-104
  sticklebacks, 170-2
  synthetic biology, 132, 144-5
  transformation phenomenon and, 41
  worms, 26, 89, 90, 96, 140, 202, 203
  see also DNA; exons; genomes; *hox* genes; *Manx* gene; minimal genome; mutations; pseudogenes; recombination; regulation; transcription factors
genetic capacitance, 120-1
genetic circuits, 91, 129, 133, 134-5, 173
  see also genetic modules; genetic networks; hotspots
genetic information, 16, 141, 161-8, 182
  chromosomes, 34-7
  DNA sequences, 11-12, 46, 48-50, 107, 128-9, 162-9, 179
  histones and epigenotypes, 100
  non-coding DNA sequences, 18, 107, 112-13, 165
  see also DNA sequences; heredity; mutations; recombination; transformation
genetic modification *see* genetic reprogramming
genetic modules, 137-41, 143
  see also genetic circuits
genetic networks, 71, 129, 138, 144
  see also genetic circuits; genetic modules
genetic reprogramming, 107-24, 171, 183
  behavioural effects, 204-6
  Escherichia coli, 134-5, 158
  human evolution, 197-9
  non-coding DNA sequences' role, 18, 107-16, 170-1
  see also synthetic biology
genetics, origin of term, 34
genome sequencing *see* DNA sequencing
genome synthesis, 133, 139, 141
  see also synthetic biology
genomes
  definition and description, 19, 79, 106-7
  see also DNA sequences
genomic churning, 115-16
genomic imprinting, 98-101
genomic maps, 82
Gibbs, Richard, 88
Gilbert, Walter, 79, 80, 106

INDEX

glutathionylation, 103
glycosylation, 103
Goethe, Johann Wolfgang von, 64
gorillas, 182, 187
Gould, Stephen Jay, 121, 154, 168
gradualistic evolution, 65-7, 119, 121, 123-4, 175
Grant, Ulysses S., 182
Greenwood, James, 9-10
Gregory (medieval author), 55
Griffiths, Fred, 38-9, 40
Guevara, Che, 181
Gunturkun, Onur, 199

haemoglobin gene, 141
Haemophilus influenzae, 84, 85, 130
Halder, Georg, 96
hammerhead ribozyme, 174-5
Han, Jeffrey, 112
haploid genes, 98
Harvey, William, 60-1, 62
Hauser, Marc D., 190
Healy, Bernadine, 83
heat-shock protein, 120-1
heredity, 31-46, 65, 108
  see also evolution
Hershey, Alfred, 42-4, 45
Hildebrandt, Nora, 53
Hippocrates, 31
histones, 99-100, 101
historical constraints, 158-61
Hitchcock, Alfred, 76
Hogness, David, 70
homeobox, 70
homeotic variations, 68-71
homunculi, 61, 62, 63
honeybees, 68-9, 142, 143
Hopkins, Nancy, 74
hotspots, 171-2, 173, 174
hox genes, 70-1, 97, 111-12, 171
  see also master control genes
hox paradox, 71
HSP90 (heat-shock protein), 120-1
human genome
  comparison with others, 26, 82, 138, 182-3, 186-8, 192, 197-8
  conserved non-genic sequences, 115
  genes, 88-90, 140, 141, 182, 183
  inference of characteristics from, 162
  microRNAs, 97
  non-coding DNA sequences, 81, 107
  nucleotides, 80
  retrotransposons, 107, 111
  sequencing, 79-88, 99, 106, 107, 182, 186
  transcription factors, 96, 97
  transference to other species, 185
Human Genome Project, 81, 106, 136
humans
  ancient DNA, 12-13, 14, 19-20
  cerebral cortex, 192-4
  jaw muscles and skull size, 195-7
  respiratory tract, 154-5
  sense of smell, 153
  see also brains
Hume, David, 205
Hutchinson, Clyde, 130
hydroxylation, 103
hyperthermophilic bacteria, 157

Ice Man, 14
imaginal disks, 95
immune systems, vertebrates, 104
immunoglobin supergene family, 140-1
imprinted/inactivated genes, 98-101
indels *see* deletions; insertions
inevitable constraints, 153-4
information *see* cultural inheritance; extragenetic information storage; genetic information
inheritance *see* cultural inheritance; genes, inheritance of; heredity
insertions, 111, 186-7
intelligence, 185, 200
  see also brains; mind
interactomes, 135
introns, 102, 112

Jacob, François, 49
James I, King, 60
JAW (microRNA molecule), 97
jaw muscles, 196-7

INDEX

Jeffrey, William, 118
Jeffries, Ann, 147
Jo Jo, 53
Johannsen, Wilhelm, 36
junk DNA sequences *see* non-coding DNA sequences

KE family, 190-1
Kellis, Manolis, 115
Kepler, Johannes, 30
khipu, 177-9, 180
Kim, Stuart, 137
Kirby, William, 66
Kirk, Revd. Robert, 147
Koonin, Eugene, 130, 131
Kornberg, Arthur, 45
Krontiris, Theodore, 113
Kwak, Shin, 102

Lamarck, Jean-Baptiste de, 65
Lander, Eric, 115, 186
language, 12, 23, 180-1, 188-92, 194
 see also cultural inheritance
Leibler, Stanislas, 144
Lenski, Richard, 124
Leonardo of Pisa, 151
lesser snow goose, 174
Lewis, Edward, 69-70
Lewontin, Richard, 154
Li, Wen-Hsiung, 114
Liddell, Alice, 14
Lindquist, Susan, 120, 121
LINEs (long interspersed elements), 110-11
Linnaean classification, 122-3
Liu, Haibo, 156
Locke, L. Leland, 178
LUCA (last universal common ancestor), 27-8
Lucretius, 57
Luria, Salvador, 43

McCarty, Maclyn, 41
McClintock, Barbara, 107, 108-9, 110
McGinnis, William, 70
MacLeod, Colin, 41

macroevolution, 119, 121, 170, 174
maize, 46-7, 107-9, 116-18
malaria, 141, 203
Malebranche, Nicolas, 62
Malpighi, Marcello, 61
Manx gene, 118, 119
Marliere, Philippe, 157
marsupials, 15
master control genes, 70, 71, 118-19
 see also hox genes
material constraints, 155-8
Mayhew, Henry, 105, 106
MeCP2 protein, 100
Mendel, Gregor, 31-4, 35, 36, 108
mental phenomena, 183-4, 186, 198, 200-1, 204
 see also brains; consciousness; mind
meristic variations, 68
Merrick, James, 53
messenger RNA *see* mRNA
metabolic networks, 18, 140, 143-4
Methanosarcina barkeri, 157
methylation, 99-101, 103, 166
mice
 brains, 192, 193
 DNA sequences, 19, 82, 84, 88, 187-8
 genes, 90, 97, 140, 201, 202
 phylum, 26
 pneumococci and, 39, 41
 retrotransposons, 107, 111
 sense of smell, 153
microevolution, 119, 170, 174
microsatellites, 113
mind, 77-8, 184, 188-9, 200
 see also brains; consciousness; mental phenomena
minimal genome, 130-1, 161-7
miRNAs (microRNAs), 89, 96-7, 101-2, 114, 115, 143, 197
mitochondria and mitochondrial DNA, 16, 17, 19, 23, 132
Model Cell Consortium, 136
modification *see* genetic reprogramming; genomic imprinting; mutations;

INDEX

recombination; synthetic biology
Monaco, Anthony, 191
Monod, Jacques, 49
Montaigne, Michel de, 78
Moore, Jeffrey, 156
Morgan, Thomas Hunt, 34-7
morphogenetic fields, 71
Morris, Simon Conway, 169
mosquitoes, 88, 203-4
mRNA (messenger RNA), 49, 92, 102-3, 115
   DNA sequencing, 83
   gene array technology, 142
   miRNAs (microRNAs) and, 97
   pseudogenes, 114
   transposons and, 112
Muller, Hermann, 92
Muller, Johannes, 127
Mundy, Nicholas, 173, 174
Mushegian, Arcady, 130, 131
mutant forms, 51-74
mutations, 16-17, 37, 42, 54
   buffering, 114, 119-21
   cerebral cortex, 193-4
   chimpanzee genes, 187
   chromosome damage and, 108-10
   developmental 'hotspots', 171-2
   duplicate genes, 114
   fixation, 24, 46, 50, 120, 171, 173
   jaw muscles, 196-7
   language disorders, 191-2
   maize, 108-9, 116-18
   microsatellite sequences, 113
   nucleotides, 171
   parallel evolution, 173-4
   position changes, 92
   preferential accumulation, 172
   retrotransposons and, 111-12
   spontaneous, 49-50, 54
   see also evolvability; homeotic variations; point mutations; post-translational control and modification; recombination
   myosin heavy chain (MYH) proteins, 196-7

N value paradox, 89

Nägeli, Carl Wilhelm von, 33
Napp, Cyrill Franz, 32
natural selection *see* Darwinism
Naturphilosophie movement, 64-5
ncRNA (non-coding RNA) genes, 115
Neanderthals, 19-20
neddylation, 103
Nelson, Laura, 202
nematode worms *see* worms
networks, 18, 71, 129, 135, 138, 140, 143-4
neurotransmitters, 199, 201
Newton, Isaac, 30, 129
Nicholas II, Tsar, 12
Nobrega, Marcelo, 113
non-coding DNA sequences, 106-16
   gene numbers and, 89
   genetic information storage, 18, 107, 112-13, 165
   genome comparisons, 92, 113-14
   human genome, 81, 106
   minimal, 130
   regulatory and reprogramming functions, 18, 107, 170-1
   shavenbaby gene, 172
   synthetic biology, 131, 133, 141
non-coding RNA (ncRNA) genes, 115
nuclear DNA, 16, 17, 23
nucleosomes, 99-100
nucleotides
   artificial DNA, 156-7
   conserved non-genic sequences (CNGs), 115
   DNA, 42, 46, 48, 179
   human genome, 80
   microsatellites, 113
   mutations and, 171
   retrotransposons, 110-11
   transposons, 112
   see also codons; DNA sequences
Nüsslein-Volhard, Christine, 72, 73

orang-utans, 167, 187, 197
orthologs, 159
ovism, 62
Owen, Sir Richard, 29-31, 50

INDEX

Paabo, Svante, 197
paintings, ancient, 20-1
pair-rule genes, 72
Palatnik, Javier, 96-7
pangenesis, 31
parallel evolution, 171-5
  see also convergence
paralogs, 159
Paré, Ambroise, 56
Patrinos, Ari, 86, 87
Pauling, Linus, 23
PCR *see* polymerase chain reaction technique
Pepper, John Henry, 75-7
Peretz, Isabelle, 182
permafrost sediments, 21-2
Phaeodactylum tricornutum, 139
phenylacetylene, 156
phosphorylation, 100, 103
phyla, 25-6, 118-19, 122-3
physical constraints, 150-1
Physiologus, 54-5, 74
Piao, Xianhua, 193, 194
pigeons, 15, 27
Pisano, Leonardo, 151
Pitt, Moses, 147
Pizarro, Francisco, 177
plants, 19, 22, 96-7, 101, 107, 200
  see also *Arabidopsis*; Fibonacci series
plumage patterns, 173-4
pneumococci, 38-41
point mutations, 16, 90, 108, 119, 174-5, 191-2
polyadenylation, 103
polymerase chain reaction technique, 17-18
  see also DNA polymerase
Pompeii, 14
Pope, Alexander, 59
'position effect' mutants, 92
post-translational control and modification, 97, 103-4
preferential accumulation, mutations, 172
preformation, 61, 62, 63
promoters
  5-HTT gene, 201-2
  complexity and, 101
  function, 93-4, 136-7
  histones and, 100
  location, 92-3
  synthetic biology, 141, 144
  transcription factors and, 95, 96, 97, 100
proteins
  complexity, 90
  DNA sequences and, 17, 18, 48-9
  families of, 159-60
  folds, 120, 159-60
  forkhead domain, 192
  function, 155
  genes and, 91-2, 97, 100-1, 102
  genetic modules and, 137
  material constraint, 155
  post-translational modifications, 103-4
  splicing, 103
  synthesis, 23-4, 25, 49
  synthetic biology, 141, 143, 157-8, 160
  viruses and, 151
  see also amino acids; beta-catenin protein; heat-shock protein; histones; myosin heavy chain; transcription factors
proteomes, 104
pseudogenes, 114, 153, 159
pufferfish, 81, 88, 140
punctuated evolution *see* discontinuous evolution; saltatory transitions

quaggas, 13-14, 15, 27
Quiring, Rebecca, 95

rats, 19, 88, 90
recombination, 16, 36-7, 108, 111
regeneration, 63
regulation
  gene expression, 91-102, 109-10, 113, 136-7, 173
  non-coding DNA sequences' role, 18, 107, 170-1
  see also enhancers; *hox* genes; master control genes; promoters

INDEX

regulatory mutations, 173
regulatory sequences, 92, 113-14, 131, 132
Reik, Wolf, 99
relative constraints, 153
replication, DNA, 16, 45-6
  see also gene duplication; retrotransposons; transcription factors
reprogramming *see* genetic reprogramming
respiratory tract, humans, 154-5
retrotransposons, 107-12
reverse transcriptase, 110
ribosomes, 49
RNA, 102-3, 110, 137, 155, 174-5, 207
  see also miRNAs (microRNAs); mRNA (messenger RNA); ncRNA (non-coding RNA) genes
Robertis, Eddy De, 70-1
Robinson, G.E., 142
rock paintings, 20-1
Romanov family, 12-13
Roosevelt, Theodore, 182
roundworm, 82, 85, 86
Rousseau, Jean-Jacques, 78
Rowe, Marvin, 21
Rubin, Gerald, 86
Rutherford, Suzanne, 120, 121

salmon, 167
saltatory transitions, 119, 121-4
  see also discontinuous evolution
Sandell, Lisa, 113
Sanger, Fred, 18, 79
sarcomeres, 196
schizophrenia, 200-1
Schleiden, Matthias, 63
Schultz, Peter, 157
Schwann, Theodore, 63
sea squirt, 88
sea urchins, 93-4, 136-7
segment-polarity genes, 72
sequencing *see* DNA sequencing
Shapiro, Michael, 170
shavenbaby gene, 172

Sheldon, Peter, 124
shotgun method, DNA sequencing, 84-5, 86
Siberia, 21-2
sickle cell mutants, 141
SINEs (short interspersed elements), 111
Sinsheimer, Robert, 79-80
skull expansion, humans, 195-7
smartness, genes, 91-104
  see also gene expression
smell, sense of, 153, 188
Smith, Hamilton, 84, 131, 132
socially acquired information *see* cultural inheritance
soul, concept of, 183-4, 185
spandrels, 154
speciation, 23-4, 109-10, 124, 174-5
  see also Darwinism; evolution
spermism, 62
splicing, 102, 103
stability, genetic systems, 46, 133-4
Stanley, Eugene, 112
Stedman, Hansell, 196, 197
Sterne, Laurence, 62
sticklebacks, 169-72
Stratton, Charles Sherwood, 51-2
Streptococcus pneumoniae, 38-41
Sturtevant, Alfred H., 92
substantive variations, 68
Sucena, Elio, 172
sulfation, 103
Sulston, John, 80, 85, 86
sumoylation, 103
Sun, Yi, 101
Swalla, Billie J., 118
Swammerdam, Jan, 61-2
synapses, 199
synthetic biology, 23-4, 129-45, 149-61, 169, 171-2, 174-6, 204-6
  see also DNA space; proteins, synthesis
Szostak, Jack, 174

telomeres, 115-16
teosinte *see* maize
Theobald, Bishop, 55

INDEX

Tomita, Masaru, 135
transcription factors, 94, 95-6, 97, 100, 101
   language disorders, 191
   transposons and, 112
   yeast, 139
   see also replication; reverse transcriptase
transformation, 37-41
   see also mutations; recombination
translation
   DNA to proteins, 48-9, 97
   see also post-translational control and modification
transpiration, 150
transposons, 107-10, 112, 141, 176
Trembley, Abraham, 62-3
Tripp, Charles, 53
Tschermak, Erich von, 34
tunicates, 118-19
Turing, Alan, 204, 205

ubiquitination, 103

Vargha-Khadem, Faraneh, 191
Vaucanson, Jacques, 127, 128
Venter, J. Craig
   human genome sequencing, 82-3, 84, 85, 86, 87
   synthetic biology, 130, 131, 132
Vesuvius, 14
Victoria, Queen, 51
viruses
   bacteria and, 37-41, 43
   DNA sequences, 19, 133, 162
   Epstein-Barr, 80
   human acquisition of genes from, 183
   mutant gene location tools, 74
   proteins and, 151
   synthetic biology, 131-2, 133
   see also bacteria; bacteriophages
vision, 153
vitalism, 37
Vries, Hugo de, 34

Waal, Francis de, 189
Wainewright, T.G., 148
Walldorf, Uwe, 95
Walpole, Spencer H., 29, 30
Walsh, Christopher, 193
Warren, Lavinia, 51-2
Watson, James, 45, 69, 81, 82, 83, 84
Wellington, Duke of, 126
whales, 168, 170, 172
White, Kevin, 142
whole-genome duplication, 115
whole-genome shotgun sequencing, 84-5, 86
Wiedemann, Hans Rudolf, 98
Wieschaus, Eric, 72
Wilkins, Maurice, 45
Willey, Rev. Mr, 52
Wilson, Fred, 53
Wimmer, Eckard, 132
Wolff, Caspar Friedrich, 64
wolves, 15
worms
   DNA sequences, 19
   gene expression, 96-7
   genes, 26, 89, 90, 140, 202, 203
   phylum, 26
   transcription factor genes, 96
   see also roundworm

Yang, James, 103
yeast
   gene duplication, 114, 115
   gene expression, 142-3
   gene inventories, 139-40
   gene methylation, 101
   genome sequencing, 82, 85
   genomic churning, 115-16
   miRNAs, 97
   mRNA molecules, 102
   transposons, 107, 110
Yekta, Soraya, 97
Yuh, Chiou-Hwa, 93

zebrafish, 73-4
zebras, 13, 15, 27